PROCEEDINGS
of the
STEKLOV INSTITUTE
of
MATHEMATICS

edited by

I. G. Petrovskiĭ and S. M. Nikol'skiĭ

number 97 (1968)

THE MILNE PROBLEM WITH ANISOTROPIC SCATTERING

by

M. V. Maslennikov

AMERICAN MATHEMATICAL SOCIETY

Providence, Rhode Island

1969

Академия Наук
Союза Советских Социалистических Республик

ТРУДЫ
ордена Ленина
МАТЕМАТИЧЕСКОГО ИНСТИТУТА
имени В. А. СТЕКЛОВА
XCVII

Ответственный редактор академик И. Г. ПЕТРОВСКИЙ
Заместитель ответственного редактора профессор С. М. НИКОЛЬСКИЙ

М. В. МАСЛЕННИКОВ

ПРОБЛЕМА МИЛНА
С АНИЗОТРОПНЫМ РАССЕЯНИЕМ

Издательство „Наука"
Москва 1968

Translated from the Russian by Andrew Yablonsky

Printed in the United States of America

PROCEEDINGS OF THE STEKLOV INSTITUTE OF MATHEMATICS
IN THE ACADEMY OF SCIENCES OF THE USSR

(Труды математического института им. В. А. Стеклова, т. XCVII, 1968)

TABLE OF CONTENTS

Page

THE MILNE PROBLEM WITH ANISOTROPIC SCATTERING

M. V. MASLENNIKOV

Preface

The Milne problem and related problems play a central role in the mathematical theory of radiation passage through matter. Thus far, however, there has been no detailed treatment of the qualitative theory of the Milne problem with sufficiently general scattering laws. The present monograph is an attempt to fill this gap.

This monograph presents the theory of the integral equations describing the diffusion of monochromatic radiation through a semi-finite plane material layer (the Milne problem). The principal physical characteristic of this process is the scattering indicatrix which determines the law of single interaction of the radiation with the material. The investigation is carried out under sufficiently simple assumptions about the functional properties of the indicatrix. All of the solutions of the Milne problem are obtained. The properties and asymptotic behavior of these solutions at large optical depths are investigated. A one-to-one correspondence between the scattering indicatrix and the directional distribution of the radiation in the interior of the layer is established.

The characteristic integral equation which relates the asymptotic characteristics of the phase density of the radiation is investigated in detail.

§1. Introduction

Let us consider the halfspace $z > 0$ of three-dimensional space. This halfspace is filled with a radiation scattering and absorbing material whose density is a function of z. Let $\psi(z,\omega) = (v/l(z))\,N(z,\omega)$, where v is the velocity of radiation propagation in the material, which we assume to be constant; $l(z)$ is the total mean free path of a particle (or quantum) at the depth z in the material; $N(z,\omega)\,d\omega$ is the volume density of the radiation propagating in the pencil of directions $d\omega$ about the direction ω (ω is a unit vector). Let $\eta(z,\omega)\,dz\,d\omega$ be the number of particles emitted per unit time by sources in the volume dz within the pencil of directions $d\omega$. Steady state radiation diffusion through the halfspace $z > 0$ is then described by the kinetic equation (see for example [1])

$$l(z)(\omega\mathbf{n})\frac{\partial\psi(z,\omega)}{\partial z}+\psi(z,\omega)=\int\limits_{\Omega}\psi(z,\omega')g(\omega\omega')d\omega'+\eta(z,\omega). \quad (1)$$

Here Ω is a unit sphere in three-dimensional space; $\mathbf{n}$ is the interior normal to the plane boundary of the halfspace $z > 0$, $\mathbf{n} \in \Omega$. The kernel of the integral operator in the right side of this equation depends on the scalar product $\omega\omega'$ of the vectors ω and ω' from Ω, and is defined by prescribing the scattering indicatrix $g(\mu)$, $\mu \in [-1, 1]$. Here $2\pi g(\mu)\, d\mu$ is the probability that the direction of the velocity of a particle will change by the angle α, $\mu < \cos\alpha < \mu + d\mu$, as a result of a single interaction with the material. Thus $g_0 = 2\pi\int_{-1}^{1} g(\mu)\, d\mu$ is the total scattering probability per event of particle-material interaction; $1 - g_0$ is the corresponding absorption probability.

Equation (1) is the local particle balance relation at the point (z, ω) of the phase space. We can add to it the boundary condition

$$\psi(0, \omega) = B(\omega) \quad \text{for } \omega\mathbf{n} > 0.$$

$B(\omega)$ is a function defined on the hemisphere $\{\omega \mid \omega \in \Omega,\ \omega\mathbf{n} > 0\}$. For convenience we extend the definition of $B(\omega)$ onto the entire sphere, setting $B(\omega) = 0$ for $\omega\mathbf{n} \leq 0$.

We assume that $l(z) > 0$ everywhere and that $\int_0^\infty l(z)^{-1} dz = \infty$. In the theory of radiation transfer the function $\tau(z) = \int_0^z l(z')^{-1} dz'$ is called the optical thickness. It has the unique inverse $z(\tau)$, $\tau \in [0, \infty)$, where $z(0) = 0$, $z(\tau) \to \infty$ as $\tau \to \infty$ and $dz(\tau)/d\tau = l(z(\tau))$. Let us set

$$\varphi(\tau, \omega) = \psi(z(\tau), \omega), \qquad \mathrm{H}(\tau, \omega) = \eta(z(\tau), \omega)$$

for $\tau > 0$. Equation (1) then becomes

$$(\omega\mathbf{n})\frac{\partial\varphi(\tau,\omega)}{\partial\tau}+\varphi(\tau,\omega)=\int\limits_{\Omega}\varphi(\tau,\omega')g(\omega\omega')d\omega'+\mathrm{H}(\tau,\omega). \quad (2)$$

The differential operator in the the left side can be inverted readily. To this end we assume that the entire analysis is being carried out in the class of functions $\phi(\tau, \omega)$ such that $\lim_{\tau=\infty} \exp(-\tau/|\omega\mathbf{n}|)\,\phi(\tau, \omega) = 0$ for $\omega\mathbf{n} < 0$. Transformation of equation (2) then yields

$$\varphi(\tau,\omega)=\hat{A}\varphi(\tau,\omega)+f(\tau,\omega). \quad (3)$$

In every case where the corresponding expressions have meaning we have

$$\hat{A}\varphi(\tau,\omega)=\begin{cases}\dfrac{1}{\omega n}\displaystyle\int_0^{\tau} e^{-\frac{\tau-\rho}{\omega n}}\,d\rho\int_{\Omega}\varphi(\rho,\omega')\,g(\omega\omega')\,d\omega' & \text{for}\quad \omega n>0,\\[2mm] \displaystyle\int_{\Omega}\varphi(\tau,\omega')\,g(\omega\omega')\,d\omega' & \text{for}\quad \omega n=0,\\[2mm] \dfrac{1}{|\omega n|}\displaystyle\int_{\tau}^{\infty} e^{-\frac{\rho-\tau}{|\omega n|}}\,d\rho\int_{\Omega}\varphi(\rho,\omega')\,g(\omega\omega')\,d\omega' & \text{for}\quad \omega n<0;\end{cases} \tag{4}$$

$$f(\tau,\omega)=B(\omega)\exp\left(-\frac{\tau}{\omega n}\right)+S(\tau,\omega);$$

$$S(\tau,\omega)=\begin{cases}\dfrac{1}{\omega n}\displaystyle\int_0^{\tau} e^{-\frac{\tau-\rho}{\omega n}}\,H(\rho,\omega)\,d\rho & \text{for}\quad \omega n>0,\\[2mm] H(\tau,\omega) & \text{for}\quad \omega n=0,\\[2mm] \dfrac{1}{|\omega n|}\displaystyle\int_{\tau}^{\infty} e^{-\frac{\rho-\tau}{|\omega n|}}\,H(\rho,\omega)\,d\rho & \text{for}\quad \omega n<0.\end{cases}$$

Further on we shall make equation (3) the object of rigorous analysis. We shall introduce classes of functions in which the definition of $\hat{A}$ in accordance with (4) is strict, investigate the existence and uniqueness of the solutions of equation (3) and of the corresponding homogeneous equation, and obtain asymptotic expansions of these solutions for large $\tau, \cdots$.

If the medium does not contain sources $(H = 0)$ and if no radiation falls on the boundary of the halfspace $z > 0$ $(B = 0)$, then (3) becomes the homogeneous equation

$$\varphi(\tau,\omega)=\hat{A}\varphi(\tau,\omega). \tag{3$'$}$$

The Milne problem is usually understood to be the problem of finding the nontrivial solutions of equation (3'). We shall extend the term "Milne problem" to cover the general equation (3), referring to problem (3') as the "homogeneous Milne problem". The term "inhomogeneous Milne problem" will be applied to problem (3) with $f \neq 0$.

The Milne problem has already been studied extensively in the simplest case of isotropic scattering $(g(\mu) \equiv \text{const})$. The mathematical aspect of this problem is treated in detail in Hopf's fundamental study [2], in a paper by Placzek and Seidel [3], and in monographs by Kourganoff [4] and Busbridge [5] (see also [6,1] and other works).

The case where $g_0 = 1$, $f = 0$ and where the indicatrix depends linearly on the cosine of the scattering angle μ is readily reducible to the above case. The appropriate construction was carried out by Mark [7]. If the indicatrix $g(\mu)$ is a polynomial of finite degree in μ, we can reduce (3′) to some integral equation in the zero moment $\phi_0(\tau) = \int_\Omega \phi(\tau, \omega)\, d\omega$ of the unknown function $\phi(\tau, \omega)$. The kernel of this equation depends on the difference between the arguments, so that the problem can be investigated by the Hopf-Wiener-Fok method [2,8-10,...]. Moreover, $\phi(\tau, \omega)$ can be expressed in terms of $\phi_0(\tau)$ in this case. These devices have been used by several authors. The most thorough published papers on the subject are those of Kuščer [11] and Davison [1]. Similar constructions were carried out by Romanova [12], Bouvier [13], and Bat′ and Zareckiĭ [14]. Other methods of solving the Milne problem with a polynomial indicatrix have been in intensive development in recent years. These include the method of K. M. Case and methods stemming from the theory of singular integral equations (of the numerous papers on the subject see, for example, [15-18]).

The situation becomes more complex when we turn to indicatrices of more general form. Here it is no longer possible to reduce (3′) to a simpler equation and to use the Wiener-Hopf method. This general problem was first investigated by Hopf, who proved [2] the existence of the conservative ($g_0 = 1$) homogeneous Milne problem with an even indicatrix $g(\mu) = g(-\mu)$.

The next important step was taken by Ambarcumjan [18], who formulated a hypothesis on the separation of variables in the solution of the homogeneous Milne problem at large optical depths, $\phi(\tau, \omega) \approx e^{k\tau}\chi(\omega)$. The parameter k and the function $\chi(\omega)$ are related by an expression which can be obtained readily from (3) (for $f \equiv 0$) by neglecting asymptotically small terms:

$$(1 + k\omega n)\,\chi(\omega) = \int_\Omega g(\omega\omega')\,\chi(\omega')\,d\omega'. \tag{6}$$

Ambarcumjan's equation (6) has nontrivial solutions only for certain values of k. If $g(\mu) = \text{const}$, then these isolated k (as is clear from (6)) satisfy the transcendental equation $(g_0/2k)\ln((1+k)/(1-k)) = 1$. In the Hopf-Chandrasekhar theory the latter is known as the "characteristic equation". We shall also apply this term to integral equation (6).

As will become evident below, investigation of equation (3) reduces in large measure to the study of the properties of the solutions of equation (5) and of the corresponding inhomogeneous equation.

The nonnegative solutions of (3) and (6) are of physical interest. We shall show that the nonnegative solutions of (6) are functions of the scalar product ωn and are therefore independent of the azimuth angle. This statement was first formulated without proof by Ambarcumjan [18]. If we are interested only in such azimuthally homogeneous solutions, we can simplify (6) by integrating it over the azimuth,

$$(1+k\mu)x(\mu)=\int_{-1}^{1} g_0(\mu,\mu')x(\mu')\,d\mu', \tag{7}$$

$$g_0(\mu,\mu')=\int_0^{2\pi} g\left(\mu\mu'+\sqrt{1-\mu^2}\sqrt{1-\mu'^2}\cos\alpha\right)d\alpha.$$

If $g(\mu)$ is a polynomial, then the kernel $g_0(\mu,\mu')$ of equation (7) is degenerate and (7) can be investigated either by algebraic methods or by the method of Chandrasekhar [6, 19]. Some simple general properties of the positive solutions of (7) and of the corresponding values of k are formulated without proof in Kuščer's paper [19] and in Sobolev's monograph [20]. A detailed investigation of (7) for an arbitrary indicatrix $g \in L_2(-1, 1)$ was carried out by the present author [21]. The principal results are presented without proof in [22]. It turns out that the dependence of the spectrum of equation (7) and of the solutions of the inhomogeneous equation corresponding to (7) on the complex parameter k is of the greatest interest for the solution of the Milne problem. The properties of the solutions of (7) are also considered in [23–26, 41].

The present monograph is a detailed study of the theory of equation (3) with the following assumptions concerning the scattering indicatrix $g(\mu)$ (§3, Condition I):

a) $g(\mu)$ is a nonnegative function of the variable $\mu \in [-1, 1]$.

b) The probability of survival of a particle in a single event of interaction with the material is positive; there is no particle breeding. This hypothesis is equivalent to the inequalities

$$0 < g_0 \le 1.$$

c) $g(\mu)$ is quadratically summable over the variable μ on the segment $[-1, 1]$.

We are interested in all the solutions of equation (3), including the alternating-sign and azimuthally inhomogeneous solutions.

The study consists of three parts: 1) the theory of the Ambarcumjan characteristic equation (§§2–6); 2) the Milne problem (§§7–11); 3) some properties of the solutions of the Milne problem in the cases of high and low absorptance of the medium, and the inverse problem of the asymptotic theory of radiation transfer (§§12 and 13).

The first part begins with §2, where information about integration on a sphere and on the properties of spherical functions which will be needed further on is summarized in several lemmas (largely without proof).

§3 concerns the properties of the integral operator $\hat{g}$,

$$\hat{g}f(\omega) = \int_{\Omega} g(\omega\omega') f(\omega') d\omega',$$

whose kernel is determined by the scattering indicatrix g.

§§4–6 form the core of the first part. Here we show that there exists a nonempty finite or countable set $\mathfrak{N}$ of values of the parameter k for which characteristic equation (6) admits of nontrivial solutions. We then proceed to establish some properties of $\mathfrak{N}$ and of the associated eigenfunctions.

Some spectral properties of the characteristic equation for fixed values of the parameter k are investigated; the dependence of the solutions of the characteristic equation (of both the homogeneous and the corresponding inhomogeneous equations) on this parameter is analyzed.

The information developed in the first part of the study is then used to construct and analyze the properties of the solutions of the Milne problem.

In the second part we isolate classes of functions of the variables τ and ω in which the solutions of the Milne problem as such are to be investigated; the properties of the operator $\hat{A}$ (see formula (4)) which acts in these classes are analyzed (§§7 and 8). The method of successive approximations is used to construct a certain specific system of solutions of the homogeneous Milne problem (§9); it is shown that every solution of the homogeneous Milne problem which belongs to a sufficiently broad class of functions is a linear combination of a finite number of elements of this system (§11). The properties of positive solutions having clear physical meaning are investigated in greater detail (§9).

The central subject of the third part of the study (§§12 and 13) is the inverse problem of reconstructing the scattering indicatrix from the angular distribution of radiation in the interior of the layer. This problem turns out to be uniquely solvable if and only if the medium absorbs radiation (§13). Arrival at this result requires certain information on the properties of the positive solution of the characteristic equation for low radiation absorptance by the medium (§12). This one-to-one correspondence between the indicatrix and asymptotic radiation in the medium in turn becomes explicit in the case of high absorptances (§12). (We note here that the asymptotic characteristics of the solution of the homogeneous Milne problem are continuous functions of the scattering indicatrix [42].)

All of the proofs in the first two parts of the study (§§2–11) are detailed. Some of the statements in §§12 and 13 are given without proof but with references to our previous papers.

Among the devices employed in the study are the analytic theory of continued fractions (§13) and certain methods of functional analysis[1] (the theory of analytic perturbations of bounded operators, the theory of cosines in Banach spaces, Laplace transformation of abstract functions of a real variable).

The principal results of the monograph have already been published in the notes [43, 44].

§2. Ancillary constructions

Let R_3 be a three-dimensional real Euclidean space with the scalar product $\xi_1\xi_2$ and with the vector length $|\xi_1| = (\xi_1\xi_1)^{1/2}$, $\xi_1, \xi_2 \in R_3$. We denote by Ω the unit sphere in R_3 and by Δ a closed rectangle on the plane (μ, ϕ): $\Delta = \{(\mu, \phi) \mid -1 \leq \mu \leq 1,\ 0 \leq \phi \leq 2\pi\}$.

Let $\sigma = \{e_1, e_2, e_3\}$ be an arbitrary orthonormal basis of R_3.[2] We associate this basis with the function ψ_σ according to the rule that if $(\mu, \phi) \in \Delta$, then $\psi_\sigma(\mu, \phi) = \sqrt{1-\mu^2}\cos\phi e_1 + \sqrt{1-\mu^2}\sin\phi e_2 + \mu e_3$. It is clear that ψ_σ maps Δ onto Ω. Let M_σ be the cross-section of Ω along the zero meridian, and let Δ_0 be the interior of Δ:

$$M_\sigma = \{\omega \mid \omega = \sqrt{1-\mu^2}\, e_1 + \mu e_3,\ \mu \in [-1, 1]\},$$
$$\Delta_0 = \{(\mu, \varphi) \mid -1 < \mu < 1,\ 0 < \varphi < 2\pi\}.$$

Now, as is easy to verify,

$$\psi_\sigma(\Delta) = \Omega,\quad \psi_\sigma(\Delta \setminus \Delta_0) = M_\sigma,$$

$$\psi_\sigma(\Delta_0) = \Omega \setminus M_\sigma,\qquad \psi_\sigma^{-1}(M_\sigma) = \Delta \setminus \Delta_0,\qquad \psi_\sigma^{-1}(\Omega \setminus M_\sigma) = \Delta_0.$$

Here ψ_σ turns out to be a continuous mapping of Δ onto Ω and a homeomorphism of Δ_0 onto $\Omega \setminus M_\sigma$. If $\omega \in \Omega \setminus M_\sigma$, then the complete original $\psi_\sigma^{-1}(\omega)$ of the vector ω with respect to the mapping ψ_σ consists of a single point $\psi_\sigma^{-1}(\omega) \subset \Delta_0$. On the other hand, if $\omega \in M_\sigma$, then there exists exactly one $\mu_0 \in [-1, 1]$ such that $\omega = \sqrt{1-\mu_0^2}\, e_1 + \mu_0 e_3$, where

$$\psi_\sigma^{-1}(\omega) = \begin{cases} \{(\mu, \varphi) \mid \mu = \mu_0,\ \varphi \in [0, 2\pi]\}, & \text{if } \quad \mu_0| = 1, \\ \{(\mu_0, 0),\ (\mu_0, 2\pi)\}, & \text{if } \quad |\mu_0| < 1. \end{cases}$$

Further on we shall be obliged to deal with numerical functions defined on Ω.

---

1) However, preference is given to simpler techniques when their use does not lengthen the formulations excessively.

2) To shorten our formulations we shall from now on consider the bases of R_3 for some single, always fixed orientation.

We assume that the measure and the Lebesgue integral (symbol: $\int_{\Omega} \cdots d\omega$) are defined on Ω in the usual way. For convenient reference we can summarize some of the properties of integration on Ω in the form of a lemma.

Lemma 1. [i] *Let $E \subset \Omega$ and let σ be an arbitrary orthonormal basis of R_3. If E is measurable on Ω, then $\psi_\sigma^{-1}(E)$ is Lebesgue-measurable on the (μ, ϕ)-plane, where* $\operatorname{mes} \psi_\sigma^{-1}(E) = \operatorname{mes}_{\Omega} E$. *Conversely, if $\psi_\sigma^{-1}(E)$ is measurable on the (μ, ϕ)-plane for some basis σ, then E is measurable on Ω.*

[ii] *Let $f(\omega)$ be a function defined on Ω, and let σ be an arbitrary orthonormal basis of R_3; let $u(\mu, \phi) = f(\psi_\sigma(\mu, \phi))$, $(\mu, \phi) \in \Delta$. The function $f(\omega)$ is measurable (summable) on Ω if and only if $u(\mu, \phi)$ is measurable (summable) on Δ. If $f(\omega)$ is summable on Ω, then*

$$\int_{\Omega} f(\omega)\, d\omega = \int_{\Delta} u(\mu, \varphi)\, d\mu\, d\varphi.$$

[iii] *Let the function $q(\mu)$ be defined and summable on the segment $[-1, 1]$, and let ω_0 be an arbitrary fixed vector from Ω. The function $q(\omega\omega_0)$ of the variable $\omega \in \Omega$ is then summable on Ω, and*

$$\int_{\Omega} q(\omega\omega_0)\, d\omega = 2\pi \int_{-1}^{1} q(\mu)\, d\mu.$$

[iv] *Let $f(\omega)$ be a function summable on Ω, let S be a rotation of R_3, and let $f_1(\omega) = f(-\omega)$, $f_2(\omega) = f(S\omega)$, $\omega \in \Omega$. Then $f_1(\omega)$ and $f_2(\omega)$ are summable on Ω, and*

$$\int_{\Omega} f_1(\omega)\, d\omega = \int_{\Omega} f_2(\omega)\, d\omega = \int_{\Omega} f(\omega)\, d\omega.$$

Lemma 1 can be proved by the ordinary methods of the theory of functions of a real variable.

Introduction of an integral on the sphere Ω enables us to construct the space $L_2(\Omega)$. As usual, $L_2(\Omega)$ contains all of the complex-valued functions $f(\omega)$ measurable on Ω for which $\int_{\Omega} |f(\omega)|^2 d\omega < \infty$. Addition and multiplication by complex numbers and the scalar products of the functions f_1 and f_2 are defined naturally in the class of these functions: $(f_1, f_2) = \int_{\Omega} f_1(\omega) \overline{f_2(\omega)}\, d\omega$ (the bar denotes the complex conjugate). Relative to these operations, $L_2(\Omega)$ turns out to be a Hilbert space with the norm $\|f\| = (f, f)^{1/2}$, $f \in L_2(\Omega)$, and the zero ϑ, $\vartheta(\omega) \equiv 0$. The collection $C(\Omega)$ of all the functions continuous on Ω forms a linear manifold in $L_2(\Omega)$. $C(\Omega)$ is a Banach space relative to the norm $\|f\|_{C(\Omega)} = \sup_{\omega \in \Omega} |f(\omega)|$. For $f \in C(\Omega)$ we have $\|f\| \leq 2\sqrt{\pi}\, \|f\|_{C(\Omega)}$.

It will be convenient for us to isolate the bases of $L_2(\Omega)$ generated by the system of spherical functions of the variables (μ, ϕ). Let

$$Y_n^{(m)}(\mu, \varphi) = \left[\frac{2n+1}{4\pi}\frac{(n-|m|)!}{(n+|m|)!}\right]^{1/2} P_n^{(m)}(\mu)\, e^{im\varphi}, \tag{8}$$
$$n = 0, 1, 2, \ldots, \qquad -n \leqslant m \leqslant n$$

for $(\mu, \phi) \in \Delta$. Here $P_n^{(m)}(\mu)$ are adjoint Legendre functions,

$$P_n^{(m)}(\mu) = (1-\mu^2)^{|m|/2}\frac{d^{|m|}}{d\mu^{|m|}}P_n(\mu), \quad n \geqslant 0, \quad |m| \leqslant n,$$

and $P_n(\mu)$ is the nth Legendre polynomial

$$P_n(\mu) = \frac{1}{2^n n!}\frac{d^n}{d\mu^n}(\mu^2-1)^n, \qquad n \geqslant 0.$$

Spherical functions (8) are orthonormalized in such a way that

$$\int_\Delta Y_n^{(m)}(\mu, \varphi)\overline{Y_{n'}^{(m')}(\mu, \varphi)}\, d\mu\, d\varphi = \delta_{nn'}\delta_{mm'}$$

for $n, n' = 0, 1, 2, \cdots, -n \leq m \leq n, -n' \leq m' \leq n'$. The system $\{Y_n^{(m)}(\mu, \phi) \mid n = 0, 1, 2, \cdots, -n \leq m \leq n\}$ is complete in $L_2(\Omega)$.

Lemma 2. *Let $\sigma = \{\mathbf{e}_1, \mathbf{e}_2, \mathbf{e}_3\}$ be an orthonormal basis of R_3, $\omega \in \Omega$. We choose a $(\mu, \phi) \in \psi_\sigma^{-1}(\omega)$ and set $Y_{\sigma n}^{(m)}(\omega) = Y_n^{(m)}(\mu, \phi)$, $n = 0, 1, 2, \cdots, -n \leq m \leq n$. The definition of $Y_{\sigma n}^{(m)}(\omega)$ does not depend on the randomly chosen $(\mu, \phi) \in \psi_\sigma^{-1}(\omega)$. It turns out that*

[i]
$$Y_{\sigma n}^{(m)} \in C(\Omega), \quad (Y_{\sigma n}^{(m)}, Y_{\sigma n'}^{(m')}) = \delta_{nn'}\delta_{mm'},$$

and that

$$Y_{\sigma n}^{(m)}(\psi_\sigma(\mu, \varphi)) = Y_n^{(m)}(\mu, \varphi), \quad n \geqslant 0,\ n' \geqslant 0,\ -n \leqslant m \leqslant n,\ -n' \leqslant m' \leqslant n'$$

for $(\mu, \phi) \in \Delta$.

[ii] *System $\{Y_{\sigma n}^{(m)} \mid n = 0, 1, 2, \cdots, -n \leq m \leq n\}$ is complete in $L_2(\Omega)$. If $f \in L_2(\Omega)$, then*

$$\|f\|^2 = \sum_{n=0}^{\infty}\sum_{m=-n}^{n}|(f, Y_{\sigma n}^{(m)})|^2$$

and

$$f = \sum_{n=0}^{\infty}\sum_{m=-n}^{n}(f, Y_{\sigma n}^{(m)})Y_{\sigma n}^{(m)}$$

(convergence is in the metric of $L_2(\Omega)$).

[iii] *For $\omega_{1,2} \in \Omega$ we have*

$$P_n(\omega_1\omega_2) = \frac{4\pi}{2n+1}\sum_{m=-1}^{n} Y_{\sigma n}^{(m)}(\omega_1)\overline{Y_{\sigma n}^{(m)}(\omega_2)}.$$

[iv] *Let S be a rotation in R_3 and let $n \geq 0$. Then*

$$Y_{\sigma n}^{(m)}(S\omega) = \sum_{k=-n}^{n} \gamma_{mk} Y_{\sigma n}^{(k)}(\omega)$$

for all m, $-n \le m \le n$ *and* $\omega \in \Omega$.

The matrix of coefficients $\Gamma = (\gamma_{mk})$ *of order* $2n+1$ *is orthonormal*: $\Gamma\Gamma^* = I$, $(\Gamma^*)_{mk} = \bar{\gamma}_{km}$.

[v] *Let* $\sigma' = \{\mathbf{e}_1', \mathbf{e}_2', \mathbf{e}_3'\}$ *be yet another orthonormal basis of* R_3, *and let* $\mathbf{e}_3' = \mathbf{e}_3$. *There then exists a* $\phi_0 \in [0, 2\pi]$ *such that*

$$Y_{\sigma' n}^{(m)}(\omega) = e^{-im\varphi_0} Y_{\sigma n}^{(m)}(\omega), \qquad \omega \in \Omega$$

for all $n = 0, 1, 2, \cdots$; $-n \le m \le n$.

The properties of spherical functions collected in Lemma 2 are either well established (see for example [27-29]) or are elementary consequences of familiar facts and of the definitions of $Y_{\sigma n}^{(m)}$ and ψ_σ.

Lemma 3. *Let us choose an arbitrary vector* $\mathbf{n} \in \Omega$ *and let us stipulate that* σ *is an arbitrary orthonormal basis with the third unit vector* $\mathbf{n}$, *i.e. that* $\sigma = \{\mathbf{e}_1, \mathbf{e}_2, \mathbf{n}\}$. *Let* s *be an integer. For each function* $f \in L_2(\Omega)$ *we set*

$$\hat{P}_{\mathbf{n}}^{(s)} f = \sum_{n=|s|}^{\infty} (f, Y_{\sigma n}^{(s)}) Y_{\sigma n}^{(s)}.$$

The series in the right side converges in the metric of $L_2(\Omega)$ *and its sum does not depend on the random choice of the basis* σ *with the third unit vector* $\mathbf{n}$. *Thus the operators* $\hat{P}_{\mathbf{n}}^{(s)}$ *(where* s *is an arbitrary number) are defined uniquely on all of* $L_2(\Omega)$, $\hat{P}_{\mathbf{n}}^{(s)}(L_2(\Omega)) \subset L_2(\Omega)$. *They have the following properties*:

[i] $\hat{P}_{\mathbf{n}}^{(s)}$ *is a projective operator.*

[ii] $\hat{P}_{\mathbf{n}}^{(s)} \hat{P}_{\mathbf{n}}^{(s')} = \delta_{ss'} \hat{P}_{\mathbf{n}}^{(s)}$ *for all pairs of integers* (s, s').

[iii] $f = \sum_{s=-\infty}^{\infty} \hat{P}_{\mathbf{n}}^{(s)} f$ *for* $f \in L_2(\Omega)$.

[iv] $\hat{P}_{\mathbf{n}}^{(s)}(C(\Omega)) \subset C(\Omega)$.

[v] $\hat{P}_{\mathbf{n}}^{(s)} Y_{\sigma n}^{(m)} = \delta_{ms} Y_{\sigma n}^{(m)}$, $n = 0, 1, 2, \cdots$, $-n \le m \le n$.

[vi] *Let* $a(\mu)$ *be a function continuous on* $[-1, 1]$:

$$f \in L_2(\Omega), \qquad h(\omega) = a(\omega\mathbf{n}) f(\omega).$$

Then

$$h \in L_2(\Omega) \text{ and } \hat{P}_{\mathbf{n}}^{(s)} h(\omega) = a(\omega\mathbf{n}) \hat{P}_{\mathbf{n}}^{(s)} f(\omega).$$

[vii] *Let* $f \in C(\Omega)$. *Then*

$$\hat{P}_{\mathbf{n}}^{(s)} f(\omega) = \frac{1}{2\pi} e^{is\varphi} \int_0^{2\pi} e^{-is\varphi'} f(\psi_\sigma(\mu, \varphi'))\, d\varphi'.$$

Here σ is an arbitrary orthonormal basis of R_3 with the third unit vector $\mathbf{n}$, *and $(\mu, \phi) \in \psi_\sigma^{-1}(\omega)$. The integral in the right side does not depend either on the randomly chosen σ with the third unit vector* $\mathbf{n}$ *or on the point $(\mu, \phi) \in \psi_\sigma^{-1}(\omega)$.*

Lemma 3 enables us to decompose every function $f \in L_2(\Omega)$ into the sum of its azimuthal moments $\hat{P}_{\mathbf{n}}^{(s)} f$ (relative to the fixed polar vector $\mathbf{n}$). To prove it we note first of all that the convergence of the series in the definition of $P_{\mathbf{n}}^{(s)} f$ follows from Lemma 2[ii]. Let σ' be another orthonormal basis with the third unit vector $\mathbf{n}$. According to Lemma 2[v] there exists a $\phi_0 \in [0, 2\pi]$ such that $Y_{\sigma' n}^{(s)}(\omega) = e^{-is\phi_0} Y_{\sigma n}^{(s)}(\omega)$. Hence

$$(f, Y_{\sigma' n}^{(s)}) Y_{\sigma' n}^{(s)} = (f, e^{-is\varphi_0} Y_{\sigma n}^{(s)}) e^{-is\varphi_0} Y_{\sigma n}^{(s)} = (f, Y_{\sigma n}^{(s)}) Y_{\sigma n}^{(s)}$$

and the definition of $\hat{P}_{\mathbf{n}}^{(s)}$ does not, in fact, depend on the chosen basis σ with the third unit vector $\mathbf{n}$.

The linearity of $\hat{P}_{\mathbf{n}}^{(s)}$ is self-evident. For $f \in L_2(\Omega)$, $n' \geq 0$, $|s'| \leq n'$ we have

$$(\hat{P}_{\mathbf{n}}^{(s)} f, Y_{\sigma n'}^{(s')}) = \sum_{n=|s|}^{\infty} (f, Y_{\sigma n}^{(s)}) (Y_{\sigma n}^{(s)}, Y_{\sigma n'}^{(s')}) = \sum_{n=|s|}^{\infty} \delta_{nn'} \delta_{ss'} (f, Y_{\sigma n}^{(s)}) = \delta_{ss'} (f, Y_{\sigma n'}^{(s')}).$$

Hence

$$\|\hat{P}_{\mathbf{n}}^{(s)} f\|^2 = \sum_{n'=0}^{\infty} \sum_{s'=-n'}^{n'} |(\hat{P}_{\mathbf{n}}^{(s)} f, Y_{\sigma n'}^{(s')})|^2$$

$$= \sum_{n'=0}^{\infty} \sum_{s'=-n'}^{n'} \delta_{ss'} |(f, Y_{\sigma n'}^{(s)})|^2 = \sum_{n'=|s|}^{\infty} |(f, Y_{\sigma n'}^{(s)})|^2 \leqslant \|f\|^2,$$

which implies that $\|\hat{P}_{\mathbf{n}}^{(s)}\| \leq 1$.

Let us choose an $n \geq 0$ and an m, $|m| \leq n$. Then

$$\hat{P}_{\mathbf{n}}^{(s)} Y_{\sigma n}^{(m)} = \sum_{n'=|s|}^{\infty} (Y_{\sigma n}^{(m)}, Y_{\sigma n'}^{(s)}) Y_{\sigma n'}^{(s)} = \sum_{n'=|s|}^{\infty} \delta_{ms} \delta_{nn'} Y_{\sigma n'}^{(s)} = \delta_{ms} Y_{\sigma n}^{(m)},$$

and Property [v] has been proved. Thus, for fixed s and s' for all $n \geq 0$, $|m| \leq n$ we have

$$\hat{P}_{\mathbf{n}}^{(s)} \hat{P}_{\mathbf{n}}^{(s')} Y_{\sigma n}^{(m)} = \delta_{ms'} \hat{P}_{\mathbf{n}}^{(s)} Y_{\sigma n}^{(m)} = \delta_{ms'} \delta_{ms} Y_{\sigma n}^{(m)} = \delta_{ss'} \delta_{ms} Y_{\sigma n}^{(m)} = \delta_{ss'} \hat{P}_{\mathbf{n}}^{(s)} Y_{\sigma n}^{(m)}.$$

Recalling that $\{Y_{\sigma n}^{(m)}\}$ is a basis of $L_2(\Omega)$, we obtain [ii].

Moreover, $(\hat{P}_{\mathbf{n}}^{(s)} Y_{\sigma n}^{(m)}, Y_{\sigma n'}^{(m')}) = \delta_{ms} (Y_{\sigma n}^{(m)}, Y_{\sigma n'}^{(m')}) = \delta_{ms} \delta_{nn'} \delta_{mm'}$; $(Y_{\sigma n}^{(m)}, \hat{P}_{\mathbf{n}}^{(s)} Y_{\sigma n'}^{(m')}) = \delta_{m's} \delta_{nn'} \delta_{mm'} = \delta_{ms} \delta_{nn'} \delta_{mm'} = (\hat{P}_{\mathbf{n}}^{(s)} Y_{\sigma n}^{(m)}, Y_{\sigma n'}^{(m')})$. Since this relation is valid for all pairs $Y^{(m)}_{\sigma n}$, $Y^{(m')}_{\sigma n'}$ $(n \geq 0,\ |m| \leq n,\ n' \geq 0,\ |m'| \leq n')$, and since $\hat{P}^{(s)}_{\mathbf{n}}$ is a linear bounded operator, it follows that $(f_1, \hat{P}^{(s)}_{\mathbf{n}} f_2) = (\hat{P}^{(s)}_{\mathbf{n}} f_1, f_2)$ for all $f_{1,2} \in L_2(\Omega)$. Thus $\hat{P}^{(s)}_{\mathbf{n}}$ is a selfadjoint operator which satisfies the condition $(\hat{P}^{(s)}_{\mathbf{n}})^2 = \hat{P}^{(s)}_{\mathbf{n}}$ (compare [ii]). This implies [i].

Let $f \in L_2(\Omega)$. From the relation $\|\hat{P}^{(s)}_{\mathbf{n}} f\|^2 = \sum_{n=|s|}^{\infty} |(f, Y^{(s)}_{\sigma n})|^2$ and Lemma 2 we infer the inequality $\sum_{s=-\infty}^{\infty} \|\hat{P}^{(s)}_{\mathbf{n}} f\|^2 < \infty$. Together with [ii] this means that the series $\sum_{s=-\infty}^{\infty} \hat{P}^{(s)}_{\mathbf{n}} f = f' \in L_2(\Omega)$ converges. But $(f', Y^{(m)}_{\sigma n}) = \sum_{s=-\infty}^{\infty} (\hat{P}^{(s)}_{\mathbf{n}} f, Y^{(m)}_{\sigma n}) = \sum_{s=-\infty}^{\infty} \delta_{sm} (f, Y^{(m)}_{\sigma n}) = (f, Y^{(m)}_{\sigma n})$, which implies that $f' = f$; we have thus proved [iii].

Now let $f \in C(\Omega)$, let s be an integer, and let σ be an orthonormal basis of R_3 with the third unit vector $\mathbf{n}$, $\sigma = \{\mathbf{e}_1, \mathbf{e}_2, \mathbf{n}\}$. We set

$$v_\sigma(\mu, \varphi) = \frac{1}{2\pi} e^{is\varphi} \int_0^{2\pi} f(\psi_\sigma(\mu, \varphi')) e^{-is\varphi'} d\varphi', \qquad (\mu, \varphi) \in \Delta.$$

Next, we choose an $\omega \in \Omega$ and show that $v_\sigma(\mu, \phi)$ is independent of the chosen σ (with the third unit vector $\mathbf{n}$), and of the pair (μ, ϕ) provided that $(\mu, \phi) \in \psi_\sigma^{-1}(\omega)$. We first assume that $(\mu_1, \phi_1) \in \psi_\sigma^{-1}(\omega)$. If $\omega \in \Omega \backslash M_\sigma$, then $(\mu_1, \phi_1) = (\mu, \phi)$ and $v_\sigma(\mu_1, \phi_1) = v_\sigma(\mu, \phi)$. Let $\omega \in M_\sigma$, so that $\omega = \sqrt{1 - \mu_0^2}\,\mathbf{e}_1 + \mu_0 \mathbf{n}$, where μ_0 is uniquely defined. If $|\mu_0| = 1$, then $\mu = \mu_1 = \mu_0$, and $\psi_\sigma(\mu, \phi') = \mu_0 \mathbf{n}$. Hence

$$v_\sigma(\mu, \varphi) = \frac{1}{2\pi} e^{is\varphi} f(\mu_0 \mathbf{n}) \int_0^{2\pi} e^{-is\varphi'} d\varphi' = \delta_{s0} f(\mu_0 \mathbf{n}).$$

Similarly, $v_\sigma(\mu_1, \phi_1) = \delta_{s0} f(\mu_0 \mathbf{n}) = v_\sigma(\mu, \phi)$. On the other hand, if $|\mu_0| < 1$, then $\phi = 0$ or 2π, $\phi_1 = 0$ or 2π and $\mu = \mu_1 = \mu_0$. The equation $v_\sigma(\mu_1, \phi_1) = v_\sigma(\mu, \phi)$ is self-evident here. Thus $v_\sigma(\mu, \phi)$ does not depend on the chosen $(\mu, \phi) \in \psi_\sigma^{-1}(\omega)$ for a fixed σ.

Now let σ' be another basis with the third unit vector $\mathbf{n}$, $(\mu, \phi) \in \psi_\sigma^{-1}(\omega)$, $(\mu_1, \phi_1) \in \psi_{\sigma'}^{-1}(\omega)$. The basis σ becomes the basis σ' upon rotation about $\mathbf{n}$ by some angle $\phi_0 \in [0, 2\pi]$. Hence $\omega = \psi_{\sigma'}(\mu_1, \phi_1) = \psi_\sigma(\mu_1, \phi_1 + \phi_0 \ (\mathrm{mod}\ 2\pi))$. This means that $(\mu_1, \phi_1 + \phi_0 \ (\mathrm{mod}\ 2\pi)) \in \psi_\sigma^{-1}(\omega)$ and

$$v_\sigma(\mu, \varphi) = v_\sigma(\mu_1, \varphi_1 + \varphi_0 \ (\mathrm{mod}\ 2\pi)) = \frac{1}{2\pi} e^{is(\varphi_1+\varphi_0)} \int_0^{2\pi} f(\psi_\sigma(\mu_1, \varphi')) e^{-is\varphi'} d\varphi'$$

$$= \frac{1}{2\pi} e^{is(\varphi_1+\varphi_0)} \int_{-\varphi_0}^{2\pi-\varphi_0} f(\psi_\sigma(\mu_1, \varphi' + \varphi_0)) e^{-is(\varphi'+\varphi_0)} d\varphi' =$$

$$= \frac{1}{2\pi} e^{is\varphi_1} \int_0^{2\pi} f(\psi_\sigma(\mu_1, \varphi' + \varphi_0 \pmod{2\pi}))\, e^{-is\varphi'}\, d\varphi'$$

$$= \frac{1}{2\pi} e^{is\varphi_1} \int_0^{2\pi} f(\psi_{\sigma'}(\mu_1, \varphi'))\, e^{-is\varphi'}\, d\varphi' = v_{\sigma'}(\mu_1, \varphi_1).$$

Now let us set

$$f_{(1)}(\omega) = v_\sigma(\mu, \varphi) = \frac{1}{2\pi} e^{is\varphi} \int_0^{2\pi} f(\psi_\sigma(\mu, \varphi'))\, e^{-is\varphi'}\, d\varphi', \quad (\mu, \varphi) \in \psi_\sigma^{-1}(\omega). \tag{9}$$

for $\omega \in \Omega$.

According to the above argument $f_{(1)}(\omega)$ does not depend either on the randomly chosen basis σ with the third unit vector $\mathbf{n}$ or on the point $(\mu, \phi) \in \psi_\sigma^{-1}(\omega)$. Thus $f_{(1)}(\omega)$ is determined uniquely when we specify $\omega \in \Omega$. Making use of (9), we can establish the continuous dependence of $f_{(1)}(\omega)$ on ω over the entire sphere Ω.

The resulting correspondence $f \to f_{(1)}$ helps us to determine the operator $\hat{\mathcal{P}}^{(s)}_{\mathbf{n}}$, i.e. $\hat{\mathcal{P}}^{(s)}_{\mathbf{n}} f = f_{(1)}$ for $f \in C(\Omega)$. It is clear that $\hat{\mathcal{P}}^{(s)}_{\mathbf{n}}$ is a linear operator, $\hat{\mathcal{P}}^{(s)}_{\mathbf{n}}(C(\Omega)) \subset C(\Omega)$. It turns out that $\|\hat{\mathcal{P}}^{(s)}_{\mathbf{n}} f\| \leq \|f\|$ for all $f \in C(\Omega)$. In fact (Lemma 1),

$$\|\hat{\mathcal{P}}^{(s)}_{\mathbf{n}} f\|^2 = \|f_{(1)}\|^2 = \int_\Delta |f_{(1)}(\psi_\sigma(\mu, \varphi))|^2\, d\mu\, d\varphi$$

$$= \frac{1}{4\pi^2} \int_\Delta \left| \int_0^{2\pi} e^{-is\varphi'} f(\psi_\sigma(\mu, \varphi'))\, d\varphi' \right|^2 d\mu\, d\varphi$$

$$\leqslant \frac{1}{2\pi} \int_{-1}^{1} \left(\int_0^{2\pi} d\varphi' \int_0^{2\pi} |f(\psi_\sigma(\mu, \varphi'))|^2\, d\varphi' \right) d\mu$$

$$= \int_\Delta |f(\psi_\sigma(\mu, \varphi'))|^2\, d\mu\, d\varphi' = \int_\Omega |f(\omega)|^2\, d\omega = \|f\|^2.$$

Now let $n \geq 0$, $|m| \leq n$, and, as before, let σ be an orthonormal basis of R_3 with the third unit vector $\mathbf{n}$. Recalling that $Y^{(m)}_{\sigma n} \in C(\Omega)$ (Lemma 3), we can compute $\hat{\mathcal{P}}^{(s)}_{\mathbf{n}} Y^{(m)}_{\sigma n}$: for $(\mu, \phi) \in \psi_\sigma^{-1}(\omega)$ we have

$$\hat{\mathcal{P}}^{(s)}_{\mathbf{n}} Y^{(m)}_{\sigma n}(\omega) = \frac{1}{2\pi} e^{is\varphi} \int_0^{2\pi} e^{-is\varphi'} Y^{(m)}_{\sigma n}(\psi_\sigma(\mu, \varphi'))\, d\varphi'$$

$$= \frac{1}{2\pi} e^{is\varphi} \int_0^{2\pi} e^{-is\varphi'} Y^{(m)}_{n}(\mu, \varphi')\, d\varphi' = \delta_{ms} Y^{(m)}_{\sigma n}(\omega) = \hat{P}^{(s)}_{\mathbf{n}} Y^{(m)}_{\sigma n}(\omega).$$

This relation enables us to assert that $\hat{P}^{(s)}_{\mathbf{n}} f = \hat{\mathcal{P}}^{(s)}_{\mathbf{n}} f$ for $f \in C(\Omega)$. In fact, let $f_N = \sum_{n=0}^{N} \sum_{m=-n}^{n} (f, Y^{(m)}_{\sigma n}) Y^{(m)}_{\sigma n}$, $f \in C(\Omega)$. Clearly, $f - f_N \in C(\Omega)$. We have $\hat{\mathcal{P}}^{(s)}_{\mathbf{n}} f = \hat{\mathcal{P}}^{(s)}_{\mathbf{n}} f_N + \hat{\mathcal{P}}^{(s)}_{\mathbf{n}} (f - f_N) = \hat{P}^{(s)}_{\mathbf{n}} f_N + \hat{\mathcal{P}}^{(s)}_{\mathbf{n}} (f - f_N) = \hat{P}^{(s)}_{\mathbf{n}} f + r_N$, where, by [i] and by the foregoing statements,

$$\| r_N \| \leqslant \| \hat{P}^{(s)}_{\mathbf{n}} (f_N - f) \| + \| \hat{\mathcal{P}}^{(s)}_{\mathbf{n}} (f - f_N) \| \leqslant 2 \| f - f_N \|,$$

so that $\lim_{N=\infty} r_N = \vartheta$ and $\hat{P}^{(s)}_{\mathbf{n}} f = \hat{\mathcal{P}}^{(s)}_{\mathbf{n}} f = f_{(1)}$, $f \in C(\Omega)$. Statements [iv] and [vii] now follow from the construction and from the properties of $f_{(1)}$. It remains for us to prove [vi].

Let $a(\mu)$ be continuous on $[-1, 1]$ and let $a_+ = \sup_{\mu \in [-1,1]} |a(\mu)|$. We assume at first that $f \in C(\Omega)$ and set $h(\omega) = a(\omega\mathbf{n}) f(\omega)$. Then $h \in C(\Omega)$. Once again, we choose an orthonormal basis σ with the third unit vector $\mathbf{n}$. For $\omega \in \Omega$ and $(\mu, \phi) \in \psi_\sigma^{-1}(\omega)$ we have $\mu = \mathbf{n}\psi_\sigma(\mu, \phi) = \omega\mathbf{n}$ and

$$\hat{P}^{(s)}_{\mathbf{n}} h(\omega) = \hat{\mathcal{P}}^{(s)}_{\mathbf{n}} h(\omega) = \frac{1}{2\pi} e^{is\varphi} \int_0^{2\pi} e^{-is\varphi'} h(\psi_\sigma(\mu, \varphi'))\, d\varphi'$$

$$= \frac{1}{2\pi} e^{is\varphi} \int_0^{2\pi} e^{-is\varphi'} a(\mathbf{n}\psi_\sigma(\mu, \varphi')) f(\psi_\sigma(\mu, \varphi'))\, d\varphi'$$

$$= a(\mu) \frac{1}{2\pi} e^{is\varphi} \int_0^{2\pi} e^{-is\varphi'} f(\psi_\sigma(\mu, \varphi'))\, d\varphi' = a(\omega\mathbf{n}) \hat{\mathcal{P}}^{(s)}_{\mathbf{n}} f(\omega) = a(\omega\mathbf{n}) \hat{P}^{(s)}_{\mathbf{n}} f(\omega).$$

On the other hand, if $f \in L_2(\Omega)$, then let $f_n \in C(\Omega)$, $n = 1, 2, \cdots$, where $\lim_{n=\infty} \|f - f_n\| = 0$. We set $h(\omega) = a(\omega\mathbf{n}) f(\omega)$ and $h_n(\omega) = a(\omega\mathbf{n}) f_n(\omega)$. Then $h \in L_2(\Omega)$ and $\|h_n - h\| \leq a_+ \|f_n - f\|$. By virtue of the foregoing statements we have $\|\hat{P}^{(s)}_{\mathbf{n}} h - a(\omega\mathbf{n}) \hat{P}^{(s)}_{\mathbf{n}} f\| \leq \|\hat{P}^{(s)}_{\mathbf{n}} (h - h_n)\| + \|a(\omega\mathbf{n}) \hat{P}^{(s)}_{\mathbf{n}} (f_n - f)\| \leq 2a_+ \|f_n - f\|$, $n = 1, 2, \cdots$, which implies [vi].

§3. The operator $\hat{g}$

Here we investigate the properties of integral operators with kernels of the form $g(\omega\omega')$, $g \in L_2(-1, 1)$. By $[g]$ we denote the norm of the function $g \in L_2(-1, 1)$: $[g]^2 = \int_{-1}^{1} |g(\mu)|^2\, d\mu$.

Theorem 1. *Let $g \in L_2(-1, 1)$. Then for $\omega_0 \in \Omega$ the function $g(\omega_0\omega)$ of the variable $\omega \in \Omega$ is an element of $L_2(\Omega)$. This makes it possible to associate the function g with the operator $\hat{g}$ whose domain of definition is $L_2(\Omega)$:*

$$\hat{g} f(\omega) = \int_\Omega g(\omega\omega') f(\omega')\, d\omega', \qquad f \in L_2(\Omega).$$

The integral in the right side exists for every fixed value of $\omega \in \Omega$. The operator

$\hat{g}$ *has the following properties:*

[i] $\hat{g}f(\omega)$ *is a continuous function of* ω *for every* $f \in L_2(\Omega)$, *so that*

$$\hat{g}(L_2(\Omega)) \subset C(\Omega) \subset L_2(\Omega).$$

[ii] *Being an operator acting from* $L_2(\Omega)$ *into* $C(\Omega)$, *the operator* $\hat{g}$ *is linear and completely continuous. Its norm satisfies the inequality*

$$\|\hat{g}\|_{L_2\to C} \leqslant \sqrt{2\pi}\,[g].$$

Proof. [1] The function $h(\omega') = g(\omega\omega')$ belongs to $L_2(\Omega)$ (Lemma 1[iii]) for a fixed $\omega \in \Omega$. Hence the integral

$$\int_\Omega h(\omega') f(\omega')\,d\omega' = \hat{g}f(\omega), \qquad f \in L_2(\Omega)$$

exists. Here

$$|\hat{g}f(\omega)|^2 \leqslant \int_\Omega |g(\omega\omega')|^2\,d\omega' \int_\Omega |f(\omega')|^2\,d\omega' = \|f\|^2 \cdot 2\pi \int_{-1}^{1} |g(\mu)|^2\,d\mu$$

and

$$|\hat{g}f(\omega)| \leqslant \sqrt{2\pi}\,[g]\,\|f\|, \qquad \omega \in \Omega. \tag{10}$$

Let $g' \in L_2(-1, 1)$ and $\delta = g' - g$. Then, clearly, $\hat{g}' - \hat{g} = \hat{\delta}$, so that

$$|\hat{g}'f(\omega) - \hat{g}f(\omega)| = |\hat{\delta}f(\omega)| \leqslant \sqrt{2\pi}\,[g' - g]\,\|f\|, \qquad f \in L_2(\Omega),\ \omega \in \Omega. \tag{11}$$

This implies that if $\{h_s \mid s = 1, 2, \dots\} \subset L_2(\Omega)$ and $\lim_{s=\infty} h_s = g$ (in the metric of $L_2(-1, 1)$), then

$$\lim_{s=\infty} \hat{h}_s f(\omega) = \hat{g}f(\omega),$$

uniformly on Ω, whatever the function $f \in L_2(\Omega)$.

Let us assume now that $g(\mu)$ is continuous on $[-1, 1]$ and that $f \in L_2(\Omega)$. Let $\epsilon > 0$ and let η be a positive number such that the inequality $|g(\mu_1) - g(\mu_2)| < \epsilon(1 + (|f|, 1))^{-1}$ is fulfilled for $\mu_{1,2} \in [-1, 1]$, $|\mu_1 - \mu_2| < \eta$. For $|\omega_1 - \omega_2| < \eta$ we have $|\omega_1\omega' - \omega_2\omega'| \leq |\omega_1 - \omega_2| < \eta$ and

$$|\hat{g}f(\omega_1) - \hat{g}f(\omega_2)| \leqslant \int_\Omega |g(\omega_1\omega') - g(\omega_2\omega')|\,|f(\omega')|\,d\omega'$$
$$\leqslant \varepsilon(1 + (|f|, 1))^{-1}(|f|, 1) < \varepsilon.$$

---

1) The arguments used here are practically identical with those used by Vladimirov [29] to prove a similar theorem.

Thus, if $g(\mu)$ is a continuous function, then $\hat{g}f(\omega)$ is continuous on Ω for every $f \in L_2(\Omega)$.

Returning to the general case $g \in L_2(-1, 1)$, we choose a sequence $\{h_s\}$ of functions continuous on $[-1, 1]$ such that $\lim_{s=\infty} [h_s - g] = 0$. Let $f \in L_2(\Omega)$. Then $\hat{h}_s f(\omega)$ are continuous functions of $\omega \in \Omega$ and at the same time $\lim_{s=\infty} \hat{h}_s f(\omega) = \hat{g}f(\omega)$ uniformly in $\omega \in \Omega$. This means that $\hat{g}f(\omega)$ is always continuous, and statement [i] is proved.

Clearly, being an operator which acts from $L_2(\Omega)$ into $C(\Omega)$, the operator $\hat{g}$ is linear. Its boundedness follows from (10): $\|\hat{g}\|_{L_2 \to C} \leq \sqrt{2\pi}\,[g]$. All that remains to prove the theorem is to establish the complete continuity of $\hat{g}$.

Let A be the bounded part of $L_2(\Omega)$, let $K = \sup_{f \in A} \|f\| < \infty$, and let $A_1 = \hat{g}(A) \subset C(\Omega)$. It is clear that the class A_1 is bounded in $C(\Omega)$: $\sup_{a \in A_1} \|a\|_{C(\Omega)} \leq \sqrt{2\pi}\,[g] K < \infty$. Let us associate with $g(\mu)$ the sequence of functions $\{h_s(\mu)\}$ continuous on $[-1, 1]$ which converges to $g(\mu)$ in the metric of $L_2(-1, 1)$. Let $\epsilon > 0$. We choose our s_0 in such a way that $[g - h_{s_0}] < \epsilon' \equiv \epsilon/(12K + 1)$, and then choose an $\eta > 0$ such that $|\mu_1 - \mu_2| < \eta$ $|h_{s_0}(\mu_1) - h_{s_0}(\mu_2)| < \epsilon'$ for $\mu_{1,2} \in [-1, 1]$. Let $a \in A_1$, $\omega_{1,2} \in \Omega$ and $|\omega_1 - \omega_2| < \eta$. Moreover, $\hat{g}f = a$ for some $f \in A$. We now have (compare (11))

$$
\begin{aligned}
|a(\omega_1) - a(\omega_2)| &= |\hat{g}f(\omega_1) - \hat{g}f(\omega_2)| \leqslant |\hat{g}f(\omega_1) - \hat{h}_{s_0}f(\omega_1)| \\
&\quad + |\hat{h}_{s_0}f(\omega_2) - \hat{g}f(\omega_2)| + |\hat{h}_{s_0}f(\omega_1) - \hat{h}_{s_0}f(\omega_2)| \\
&\leqslant 2\sqrt{2\pi}\varepsilon' K + \int_\Omega |h_{s_0}(\omega_1\omega') - h_{s_0}(\omega_2\omega')|\,|f(\omega')|\,d\omega' \\
&\leqslant 2\sqrt{2\pi}\varepsilon' K + 2\sqrt{\pi}\varepsilon' K < \varepsilon.
\end{aligned}
$$

Since a is an arbitrary element of A_1, we conclude that the functions of the class A_1 are equicontinuous. The Arzelà theorem now implies the compactness of A_1 and at the same time the complete continuity of the mapping $\hat{g}$: $L_2(\Omega) \to C(\Omega)$.

Because of the fact that $\hat{g}$: $L_2(\Omega) \to C(\Omega)$, which we established in Theorem 1, we are also able (depending on the circumstances) to consider $\hat{g}$ as a mapping of $L_2(\Omega)$ into $L_2(\Omega)$, of $C(\Omega)$ into $C(\Omega)$, or of $C(\Omega)$ into $L_2(\Omega)$. In each of these cases $\hat{g}$ remains a completely continuous linear operator (this follows directly from Theorem 1 [ii]). Here, as we can readily verify,

$$
\|\hat{g}\|_{L_2 \to L_2} \leqslant 2\pi\sqrt{2}\,[g];
$$

$$
\|\hat{g}\|_{C \to C} \leqslant 2\pi \int_{-1}^{1} |g(\mu)|\,d\mu \leqslant 2\pi\sqrt{2}\,[g];
$$

$$\|\hat{g}\|_{C\to L_2} \leqslant 4\pi\sqrt{\pi}\int_{-1}^{1}|g(\mu)|\,d\mu \leqslant 4\pi\sqrt{2\pi}\,[g].$$

From now on, unless we indicate otherwise, we shall always assume that $\hat{g}$ is an operator acting from $L_2(\Omega)$ into $L_2(\Omega)$ with the norm $\|\hat{g}\| \equiv \|\hat{g}\|_{L_2\to L_2}$.

Theorem 2. *Let $g \in L_2(-1, 1)$. The function $g(\mu)$ can then be expanded in a series in Legendre polynomials which converges at least in the mean:*

$$g(\mu) = \sum_{n=0}^{\infty}\frac{2n+1}{4\pi}g_nP_n(\mu), \text{ где } g_n = 2\pi\int_{-1}^{1}g(\mu)P_n(\mu)\,d\mu,$$

$$n = 0, 1, 2, \ldots; \quad \sum_{n=0}^{\infty}\left(n+\frac{1}{2}\right)|g_n|^2 = 4\pi^2\,[g]^2.$$

Choose an arbitrary orthonormal basis σ of the space R_3. Then

[i] *The series representing the function $\hat{g}f(\omega)$, $f \in L_2(\Omega)$, namely*

$$\hat{g}f(\omega) = \sum_{n=0}^{\infty} g_n \sum_{m=-n}^{n}(f, Y_{\sigma n}^{(m)})\,Y_{\sigma n}^{(m)}(\omega)$$

converges uniformly in $\omega \in \Omega$.

[ii] *(The Funk-Hecke theorem [30, 31]). For $n = 0, 1, 2, \cdots$, $-n \leq m \leq n$, we have*

$$\hat{g}Y_{\sigma n}^{(m)} = g_nY_{\sigma n}^{(m)}.$$

Conversely, if k is an eigenvalue of the operator $\hat{g}$, then $k = g_n$ for some n, and the corresponding eigensubspace is the minimal subspace containing the system $\{Y_{\sigma n}^{(m)} \mid g_n = k\}$. This subspace is finite-dimensional for $k \neq 0$.

[iii] *Let $\mathbf{n}$ be a fixed vector from Ω and let the projective operators $\hat{P}_{\mathbf{n}}^{(s)}$ be defined in accordance with Lemma 3. The operator $\hat{g}$ then commutes with every $\hat{P}_{\mathbf{n}}^{(s)}$, $s = 0, \pm 1, \pm 2, \cdots$.*

Proof. Let us set

$$g_n = 2\pi\int_{-1}^{1}g(\mu)P_n(\mu)\,d\mu, \qquad S_N = \sum_{n=0}^{N}\frac{2n+1}{4\pi}g_nP_n, \quad \delta_N = g - S_N.$$

According to the Parseval condition for Legendre polynomials,

$$[\delta_N]^2 = [g]^2 - \sum_{n=0}^{N}\frac{2n+1}{8\pi^2}|g_n|^2 \to 0 \quad \text{as} \quad N\to\infty,$$

which implies that

$$\sum_{n=0}^{\infty}\left(n+\frac{1}{2}\right)|g_n|^2 = 4\pi^2\,[g]^2.$$

Let $f \in L_2(\Omega)$ and $\omega \in \Omega$. From (10) we infer that

$$\left| \hat{g} f(\omega) - \sum_{n=0}^{N} \frac{2n+1}{4\pi} g_n \int_{\Omega} P_n(\omega\omega') f(\omega')\, d\omega' \right| = |\hat{\delta}_N f(\omega)| \leqslant \sqrt{2\pi}\, [\delta_N] \|f\|,$$

so that

$$\hat{g} f(\omega) = \sum_{n=0}^{\infty} \frac{2n+1}{4\pi} g_n \int_{\Omega} P_n(\omega\omega') f(\omega')\, d\omega',$$

where the series in the right side converges uniformly on Ω. Applying the addition theorem for Legendre polynomials in the form [iii] (Lemma 2), we arrive at statement [i] of the theorem.

The resulting representation of $\hat{g}f(\omega)$ implies, among other things, that

$$\begin{aligned} \hat{g} Y_{\sigma n}^{(m)} &= \sum_{n'=0}^{\infty} g_{n'} \sum_{m'=-n'}^{n'} (Y_{\sigma n}^{(m)}, Y_{\sigma n'}^{(m')}) Y_{\sigma n'}^{(m')} \\ &= \sum_{n'=0}^{\infty} \delta_{nn'} g_{n'} \sum_{m'=-n'}^{n'} \delta_{mm'} Y_{\sigma n'}^{(m')} = g_n Y_{\sigma n}^{(m)}. \end{aligned}$$

Moreover,

$$(\hat{g} f, Y_{\sigma n}^{(m)}) = \sum_{n'=0}^{\infty} g_{n'} \sum_{m'=-n'}^{n'} (f, Y_{\sigma n'}^{(m')}) (Y_{\sigma n'}^{(m')}, Y_{\sigma n}^{(m)}) = g_n (f, Y_{\sigma n}^{(m)}) \tag{12}$$

for $f \in L_2(\Omega)$. Hence, if k is an eigenvalue of the operator $\hat{g}$, and if $f \in L_2(\Omega)$ is the corresponding eigenfunction, $f \neq \vartheta$, then for all $n = 0, 1, 2, \cdots$, $|m| \leq n$ we have

$$(k - g_n)(f, Y_{\sigma n}^{(m)}) = 0. \tag{13}$$

Because $\{Y_{\sigma n}^{(m)}\}$ is complete in $L_2(\Omega)$, there exists one pair of indices n, m such that $(f, Y_{\sigma n}^{(m)}) \neq 0$. In this case $k = g_n$. Let us denote by $\mathfrak{L}$ the eigensubspace of the operator g which corresponds to the eigenvalue k, and by H the minimal subspace containing $\{Y_{\sigma n}^{(m)} | g_n = k\}$. Since $\hat{g} Y_{\sigma n}^{(m)} = k Y_{\sigma n}^{(m)}$ for $g_n = k$, it follows that $\{Y_{\sigma n}^{(m)} | g_n = k\} \subset \mathfrak{L}$, so that $H \subset \mathfrak{L}$. Let $f \in \mathfrak{L}$. Then $f = f_1 + f_2$, where $f_1 \in H$, $f_2 \perp H$. We have $f_1 \in H \subset \mathfrak{L}$, so that $f_2 \in \mathfrak{L}$, which implies that $\hat{g} f_2 = k f_2$. Now, if $g_n \neq k$, then, by (13), $(f_2, Y_{n_0}^{(m_0)}) = 0$. If $g_{n_0} = k$, then $Y_{\sigma n_0}^{(m_0)} \in H \perp f_2$, and $(f_2, Y_{\sigma n_0}^{(m_0)}) = 0$. This means that $f_2 = \vartheta$, $f = f_1 \in H$ and $\mathfrak{L} = H$.

If $k \neq 0$, then the equation $g_n = k$ is valid for only a finite number of subscripts n (since $g_n \to 0$ as $n \to \infty$), and the dimensionality of H is

$$\sum_{g_n = k} (2n+1) < \infty.$$

[ii] has been proved.

To prove [iii] we choose an orthonormal basis σ of the space R_3 with the third unit vector $\mathbf{n}$. From [ii] and Lemma 3[v], for any spherical function $Y^{(m)}_{\sigma n}$ we have

$$\hat{P}^{(s)}_{\mathbf{n}}\hat{g}Y^{(m)}_{\sigma n} = g_n\hat{P}^{(s)}_{\mathbf{n}}Y^{(m)}_{\sigma n} = g_n\delta_{ms}Y^{(m)}_{\sigma n},$$
$$\hat{g}\hat{P}^{(s)}_{\mathbf{n}}Y^{(m)}_{\sigma n} = \delta_{ms}\hat{g}Y^{(m)}_{\sigma n} = g_n\delta_{ms}Y^{(m)}_{\sigma n}.$$

[iii] is now self-evident.

Lemma 4. *Let $g \in L_2(-1, 1)$ and let $p \geq 2$ be an integer. We set*

$$g_{(p)}(\mu) = \sum_{n=0}^{\infty} \frac{2n+1}{4\pi} g_n^p P_n(\mu).$$

The series in the right side converges uniformly in $\mu \in [-1, 1]$. It turns out that $\hat{g}_{(p)} = (\hat{g})^p$.

Proof. By Theorem 2, $\Sigma_{n=0}^{\infty} n|g_n|^2 \leq 4\pi^2[g]^2$. Let n_0 be such that

$$\sum_{n=0}^{\infty} n|g_n|^p \leqslant \sum_{n=0}^{n_0} n|g_n|^p + \sum_{n=n_0+1}^{\infty} n|g_n|^2 < 4\pi^2[g]^2 + \sum_{n=0}^{n_0} n|g_n|^p < \infty$$

for $n \geq n_0$, $|g_n| < 1$. This implies the uniform convergence of the series which defines $g_{(p)}(\mu)$. It is clear that $g_{(p)} \in C(-1, 1) \subset L_2(-1, 1)$. Let us associate the operator $\hat{g}_{(p)}$ with the function $g_{(p)}$. For $f \in L_2(\Omega)$ we have

$$\hat{g}_{(p)}f = \sum_{n=0}^{\infty} g_n^p \sum_{m=-n}^{n} (f, Y^{(m)}_{\sigma n}) Y^{(m)}_{\sigma n}$$

(σ is an arbitrary orthonormal basis of R_3). We have (compare (12))

$$(\hat{g})^2 f = \sum_{n=0}^{\infty} g_n \sum_{m=-n}^{n} (\hat{g}f, Y^{(m)}_{\sigma n}) Y^{(m)}_{\sigma n} = \sum_{n=0}^{\infty} g_n^2 \sum_{m=-n}^{n} (f, Y^{(m)}_{\sigma n}) Y^{(m)}_{\sigma n} = \hat{g}_{(2)}f$$

and $(\hat{g})^2 = \hat{g}_{(2)}$. The equality $(\hat{g})^p = \hat{g}_{(p)}$ for $p > 2$ can be established by induction over p.

Theorem 3. *If $g(\mu)$ is a real function, $g \in L_2(-1, 1)$, then g is a selfadjoint operator.*

To prove this it is clearly enough to establish that $(\hat{g}Y^{(m)}_{\sigma n}, Y^{(m')}_{\sigma n'}) = (Y^{(m)}_{\sigma n}, \hat{g}Y^{(m')}_{\sigma n'})$ for an arbitrary orthonormal basis σ of the space R_3 and for the arbitrary pair $Y^{(m)}_{\sigma n}$, $Y^{(m')}_{\sigma n'}$ of spherical functions. This is easy to prove by means of Theorem 2 [ii] and the realness of the coefficients g_n, $n = 0, 1, 2, \cdots$.

Until now we have been conducting our discussion under assumptions about the function g which are more general than those required in problems of radiation transfer theory. In the discussion to follow we shall confine our attention to functions g satisfying the following condition:

I. $g \in L_2(-1, 1)$; $g(\mu) \geq 0$ for all $\mu \in [-1, 1]$; $0 < 2\pi \int_{-1}^{1} g(\mu)\, d\mu \leq 1$.

The functions which satisfy Condition I (and only these functions) will be called "indicatrices".

The condition $g \in L_2(-1, 1)$ enables us to use the convenient apparatus of Hilbert spaces. However, if we were to keep strictly to the probabilistic meaning of the scattering indicatrix (§1), it would be more natural to subject g to the milder restriction $g \in L_1(-1, 1)$. The possibility of such a generalization is a consequence of the results of Vladimirov [29, 30], who proved the Funk-Hecke theorem and the complete continuity of the mapping $\hat{g}$: $L_p(\Omega) \to L_p(\Omega)$, $p \geq 1$, under the assumption $g \in L_1(-1, 1)$.

It is clear that

$$0 < g_0 \leqslant 1; \quad | g_n | < g_0, \; n = 1, 2, \ldots \tag{14}$$

for every indicatrix $g(\mu)$. This enables us to prove the following theorem.

Theorem 4. *Let $g(\mu)$ be an indicatrix. Then*

[i] $\|\hat{g}\| = g_0$.

[ii] *If $f \in L_2(\Omega)$ and $(\hat{g}f, f) = g_0(f, f)$, then $\hat{g}f = g_0 f$ and $f(\omega) \equiv$* const, $\omega \in \Omega$.

[iii] *Let $g_{(p)}(\mu)$ be defined in accordance with Lemma 4, $p \geq 2$. Then*

$$\lim_{p=\infty} g_0^{-p} g_{(p)}(\mu) = \frac{1}{4\pi}$$

uniformly in $\mu \in [-1, 1]$.

Proof. Because the operator $\hat{g}$ is completely continuous its spectrum $\sigma(\hat{g})$ is a pure point spectrum, and $\sigma(\hat{g}) = \{g_n \mid n = 0, 1, 2, \cdots\} \cup \{0\}$ (Theorem 2 [ii]). Hence (compare (14)) we have

$$\inf_{\nu \geqslant 1} \| (\hat{g})^\nu \|^{1/\nu} = \sup_{\lambda \in \sigma(\hat{g})} | \lambda | = g_0.$$

But $\hat{g}$ is a selfadjoint operator (Theorem 3), so that $\|(\hat{g})^\nu\|^{1/\nu} = \|\hat{g}\|$, $\nu = 1, 2, \cdots$. This implies [i].

Let $f \in L_2(\Omega)$ and $(\hat{g}f, f) = g_0(f, f)$. Recalling [i], we can write $(\hat{g}f, f) = \|\hat{g}\|(f, f)$. This means that $\hat{g}f = \|\hat{g}\| f = g_0 f$. From Theorem 2 [ii] and (14) we now find that $f(\omega) = C Y_{\sigma 0}^{(0)}(\omega) = C/2\sqrt{\pi}$, $C =$ const, and [ii] has been proved.

To prove [iii] we note that (as (14) implies) $\sup_{n \geq 1} |g_n| = \theta g_0$, $0 \leq \theta < 1$. For $p \geq 2$ we have

$$g_0^{-p} g_{(p)}(\mu) = \frac{1}{4\pi} + \varepsilon_p(\mu), \; \varepsilon_p(\mu) = \frac{1}{4\pi} \sum_{n=1}^{\infty} (2n+1) \left(\frac{g_n}{g_0} \right)^p P_n(\mu).$$

Let $\epsilon > 0$. There exists a natural number N such that $\sum_{n=N+1}^{\infty} (2n+1)|g_n|^2 < 2\pi\epsilon g_0^2$ (Theorem 2). In choosing this N, we pick our p_0 in such a way as to ensure that $\theta^{p_0} < 2\pi\epsilon/N(N+2)$. Then for $p > p_0$ it follows immediately that

$$|\varepsilon_p(\mu)| \leqslant \frac{1}{4\pi}\sum_{n=1}^{N}(2n+1)\theta^p + \frac{1}{4\pi g_0^2}\sum_{n=N+1}^{\infty}(2n+1)|g_n|^2 < \varepsilon$$

for all $\mu \in [-1, 1]$. Theorem 4 is proved.

§4. The characteristic equation

In this section we consider the characteristic equation

$$(1 + k\omega n)\, x(\omega) = \hat{g}x(\omega) \tag{15}$$

under the assumption that the operator $\hat{g}$ is defined by some indicatrix $g(\mu)$ subject to Condition I.

Let us choose a vector $\mathbf{n} \in \Omega$; let Z_0 be a complex plane with cuts along the real axis extending from $-\infty$ to -1 and from $+1$ to $+\infty$:

$$Z_0 = \{k \mid \inf_{\alpha\in[-1,1]} |1 + \alpha k| > 0\}.$$

For $k \in Z_0$ we set

$$\hat{U}(k) f(\omega) = \frac{1}{1 + k\omega n}\hat{g}f(\omega), \quad f \in L_2(\Omega). \tag{16}$$

The operator $\hat{U}(k)$, $k \in Z_0$, is a linear completely continuous operator which acts from $L_2(\Omega)$ into $L_2(\Omega)$, where $\hat{U}(k)(L_2(\Omega)) \subset C(\Omega)$ (Theorem 1). The spectrum $\sigma(\hat{U}(k))$ of the operator $\hat{U}(k)$ is a pure point spectrum. For $k \in (-1, 1)$ the operator $\hat{U}(k)$ is amenable to symmetrization. In fact, let

$$\hat{V}(k) f(\omega) = \sqrt{1 + k\omega n}\, \hat{U}(k) \frac{f(\omega)}{\sqrt{1 + k\omega n}}, \quad k \in (-1, 1) \tag{17}$$

for $f \in L_2(\Omega)$. It is clear that $\hat{V}(k)$ is a linear completely continuous selfadjoint operator acting in $L_2(\Omega)$. The symmetrizability of $\hat{U}(k)$ implies that for $k \in (-1, 1)$ every point $\lambda \in \sigma(\hat{U}(k))$, $\lambda \neq 0$, is a real eigenvalue of rank 1 of the operator $\hat{U}(k)$. The Fredholm theorems hold for the operator $\hat{U}(k)$. In the interests of the discussion to follow, we shall state one of these in the following form.

Lemma 5. *Let $k \in (-1, 1)$ and let λ be a number, $\lambda \neq 0$, $\alpha \in L_2(\Omega)$. The equation*

$$(\hat{U}(k) - \lambda\hat{I})\, \psi(\omega) = -\frac{\alpha(\omega)}{1 + k\omega n} \tag{18}$$

admits of a solution $\psi \in L_2(\Omega)$ if and only if $(\alpha, f) = 0$ for every function $f \in L_2(\Omega)$ which satisfies the condition $(\hat{U}(k) - \lambda\hat{I}) f = \vartheta$ ($\hat{I}$ is the identity operator, $\hat{I}f = f$).

The lemma is self-evident if $\lambda \bar{\in} \sigma(\hat{U}(\mathrm{k}))$. If $\lambda \in \sigma(\hat{U}(\mathrm{k}))$, then, as was already noted, λ is a real number. We denote by $\hat{U}_\lambda^*$ the operator adjoint to the operator $\hat{U}_\lambda \equiv \hat{U}(\mathrm{k}) - \lambda \hat{I}$. It is easy to verify that $U_\lambda^* u = \vartheta$ for $u \in L_2(\Omega)$ if and only if $\hat{U}_\lambda u(\omega)/(1 + \mathrm{k}\omega\mathbf{n}) = \vartheta$. Hence, if equation (18) admits of a solution and if $\hat{U}_\lambda f = \vartheta$, then $\hat{U}_\lambda^* u = \vartheta$, where $u(\omega) = (1 + \mathrm{k}\omega\mathbf{n})f(\omega)$. By virtue of the Fredholm theorem $(\alpha/(1 + \mathrm{k}\omega\mathbf{n}), u) = 0$, or, which is the same thing (since k is a real number), $(\alpha, f) = 0$. The inverse argument can be carried out in the same way.

Lemma 6. [i] *Let* $\mathrm{k} \in Z_0$, $f \in L_2(\Omega)$, *and let* $f_1(\omega) = f(-\omega)$ *for* $\omega \in \Omega$. *Then* $-\mathrm{k} \in Z_0$, $f_1 \in L_2(\Omega)$ *and* $\hat{U}(-\mathrm{k}) f_1(\omega) = \hat{U}(\mathrm{k}) f(-\omega)$.

[ii] *For* $\mathrm{k} \in Z_0$ *we have*

$$\sigma(\hat{U}(\mathrm{k})) \setminus \{0\} = \sigma(\hat{U}(-\mathrm{k})) \setminus \{0\}.$$

[iii] *Let* $\mathrm{k} \in (-1, 1)$, $\mathrm{k} \neq 0$, *and let* $f(\omega)$ *be a real function*, $f \in L_2(\Omega)$, $f \neq \vartheta$, *and* $\hat{U}(\mathrm{k}) f = f$. *Then* $f \in C(\Omega)$ *and* $\mathrm{k}(\omega\mathbf{n}, f^2(\omega)) < 0$.

In fact, for $\mathrm{k} \in Z_0$, $f \in L_2(\Omega)$ and $f_1(\omega) = f(-\omega)$ the inclusions $-\mathrm{k} \in Z_0$ and $f_1 \in L_2(\Omega)$ follow from the definition of Z_0 and from Lemma 1 [iv]. By virtue of the same lemma

$$\hat{U}(-\mathrm{k}) f_1(\omega) = \frac{1}{1 - \mathrm{k}\omega\mathrm{n}} \int_\Omega g(\omega\omega') f_1(\omega')\, d\omega' = \frac{1}{1 + \mathrm{k}(-\omega)\mathrm{n}} \int_\Omega g((-\omega)(-\omega'))$$

$$\times f(-\omega')\, d\omega' = \frac{1}{1 + \mathrm{k}(-\omega)\mathrm{n}} \int_\Omega g((-\omega)\omega') f(\omega')\, d\omega'$$

$$= \hat{U}(\mathrm{k}) f(-\omega) \quad (\equiv [\hat{U}(\mathrm{k}) f](-\omega)).$$

Let $\lambda \in \sigma(\hat{U}(\mathrm{k})) \setminus \{0\}$. This means that there exists an $f \in L_2(\Omega)$, $f \neq \vartheta$, such that $\hat{U}(\mathrm{k}) f = \lambda f$. Let $f_1(\omega) = f(-\omega)$, $\omega \in \Omega$. Then $\|f_1\| = \|f\| > 0$ (Lemma 1), and $\hat{U}(-\mathrm{k}) f_1(\omega) = \hat{U}(\mathrm{k}) f(-\omega) = \lambda f_1(\omega)$, so that $\lambda \in \sigma(\hat{U}(-\mathrm{k})) \setminus \{0\}$. Thus $\sigma(\hat{U}(\mathrm{k})) \setminus \{0\} \subset \sigma(\hat{U}(-\mathrm{k})) \setminus \{0\}$.

Let us accept assumptions [iii] and assume that $\mathrm{k}(\omega\mathbf{n}, f^2) \geq 0$. We have $f(\omega) + \mathrm{k}(\omega\mathbf{n}) f(\omega) = \hat{g} f(\omega)$, whence (see Theorem 4 [i] and (14)) it follows that

$$(f, f) \leqslant (f, f) + \mathrm{k}(\omega\mathbf{n}, f^2) = (\hat{g} f, f) \leqslant g_0 (f, f) \leqslant (f, f).$$

This means that $g_0 = 1$, $(\hat{g} f, f) = (f, f)$. Making use of Theorem 4 [ii], we find that $f(\omega) \equiv C = \mathrm{const} \neq 0$ and $\hat{g} f(\omega) \equiv C$. Thus $1 + \mathrm{k}\omega\mathbf{n}$ is necessarily equal to unity, which contradicts the assumption that $\mathrm{k} \neq 0$.

Definition. We denote by $\mathfrak{N}$ the set of all values of the parameter $\mathrm{k} \in Z_0$ for which characteristic equation (15) admits of nontrivial solutions.

Theorem 5. $\mathfrak{N} \subset (-1, 1)$. *The set* $\mathfrak{N}$ *is symmetric to the point* $\mathrm{k} = 0$: $-\mathfrak{N} = \mathfrak{N}$.

The set $\mathfrak{N}$ *is not more than countable. If* $\mathfrak{N}$ *is infinite, then the points* $k = 1$ *and* $k = -1$ *are the limit points of the set* $\mathfrak{N}$, *which has no other limit points.*

Proof. It is clear that $\mathfrak{N} = \{k | k \in Z_0,\ 1 \in \sigma(\hat{U}(k))\}$. This formula and Lemma 6[ii] implies that, along with $k \in \mathfrak{N}$, the set $\mathfrak{N}$ also contains the point $-k$. Thus $\mathfrak{N}$ is symmetric, i.e. $-\mathfrak{N} = \mathfrak{N}$.

Let $k \in \mathfrak{N}$. By the definition of $\mathfrak{N}$ we have $k \in Z_0$, and there exists a function $f \in L_2(\Omega)$, $f \neq \vartheta$, such that $(1 + k\omega\mathbf{n})f = \hat{g}f$. Hence $(f, f) + k((\omega\mathbf{n})f, f) = (\hat{g}f, f)$. Since $\hat{g}$ is a selfadjoint operator, we find that the product $k((\omega\mathbf{n})f, f)$ is a real number. At the same time $((\omega\mathbf{n})f, f) = \int_\Omega (\omega\mathbf{n})|f(\omega)|^2 d\omega$ is a real number. Hence k is real for $((\omega\mathbf{n})f, f) \neq 0$, so that $k \in (-1, 1)$. Let $((\omega\mathbf{n})f, f) = 0$. Then $(f, f) = (\hat{g}f, f) \leq g_0(f, f) \leq (f, f)$. This means that $g_0 = 1$ and $f(\omega) \equiv C = \text{const}$, $\omega \in \Omega$ (Theorem 4). The condition $f \neq \vartheta$ implies that $C \neq 0$. The relation $(1 + k\omega\mathbf{n})f = \hat{g}f$ now becomes $1 + k\omega\mathbf{n} = 1$, $\omega \in \Omega$, and hence $k = 0 \in (-1, 1)$. The inclusion $\mathfrak{N} \subset (-1, 1)$ has been proved.

To prove the theorem we need merely establish the finiteness of the set $\mathfrak{N} \cap [a, b]$, where a and b are any numbers satisfying the inequalities $-1 < a < b < 1$.

Let us assume the contrary, i.e. that the set $\mathfrak{N} \cap [a, b]$ is infinite for some pair of numbers a and b, $-1 < a < b < 1$. There then exist sequences $\{k_\nu | \nu = 1, 2, \cdots\} \subset \mathfrak{N} \cap [a, b]$ and $\{f_\nu | \nu = 1, 2, \cdots\} \subset L_2(\Omega)$ such that

a) the limit $\lim_{\nu=\infty} k_\nu = \lambda \in [a, b]$ exists;

b) $k_{\nu'} \neq k$ for $1 \leq \nu < \nu'$; $k_\nu \neq \lambda$, for $\nu = 1, 2, \cdots$;

c) $f_\nu(\omega)$ is a real function of $\omega \in \Omega$; $\|f_\nu\| = 1$, and

$$(1 + k_\nu\omega\mathbf{n}) f_\nu(\omega) = \hat{g}f_\nu(\omega),\ \nu = 1, 2, \ldots. \tag{19}$$

The sequence $\{f_\nu\}$ is bounded in $L_2(\Omega)$, and the operator $\hat{g}$ is completely continuous (Theorem 1). Hence $\{\hat{g}f_\nu | \nu = 1, 2, \cdots\}$ is compact in $L_2(\Omega)$. Turning to a subsequence if required, we can assume that $\{\hat{g}f_\nu\}$ converges in $L_2(\Omega)$, i.e. that $\hat{g}f_\nu \to h \in L_2(\Omega)$. We set $f(\omega) = (1/(1 + \lambda\omega\mathbf{n}))\, h(\omega)$, $\omega \in \Omega$. Clearly, $f \in L_2(\Omega)$ and

$$\|f_\nu - f\| \leqslant (1 - \max\{|a|, |b|\})^{-1} \|h - (1 + \lambda\omega\mathbf{n}) f_\nu\|$$
$$\leqslant (1 - \max\{|a|, |b|\})^{-1} [\|h - \hat{g}f_\nu\| + |k_\nu - \lambda|],\ \nu = 1, 2, \ldots.$$

Thus $\lim_{\nu=\infty} f_\nu = f$. From this it follows that $f(\omega)$ is a real function, $\|f\| = 1$. From the inequality

$$\|(1 + \lambda\omega\mathbf{n}) f - \hat{g}f\| \leqslant \|(1 + \lambda\omega\mathbf{n}) f - (1 + \lambda\omega\mathbf{n}) f_\nu\| + \|(1 + \lambda\omega\mathbf{n}) f_\nu$$
$$- (1 + k_\nu\omega\mathbf{n}) f_\nu\| + \|\hat{g}f_\nu - \hat{g}f\| \leqslant 2\|f - f_\nu\| + |k_\nu - \lambda| + g_0\|f - f_\nu\|,$$
$$\nu = 1, 2, \ldots,$$

we find that

$$(1+\lambda\omega\mathbf{n})f(\omega)=\hat{g}f(\omega),\qquad \omega\in\Omega. \tag{20}$$

This means, among other things, that $f(\omega)$ is continuous on Ω (Theorem 1). Let us take the scalar products of the right side of (20) and f_ν, and of the left side of equation (19) and f:

$$(f, f_\nu)+\lambda((\omega\mathbf{n})f, f_\nu)=(\hat{g}f, f_\nu),$$
$$(f, f_\nu)+k_\nu((\omega\mathbf{n})f, f_\nu)=(f, \hat{g}f_\nu),\qquad \nu=1, 2,\dots .$$

Because $\hat{g}$ is selfadjoint, it follows that $(k_\nu-\lambda)((\omega\mathbf{n})f, f_\nu)=0$, $\nu=1, 2,\cdots$. But $k_\nu\neq\lambda$, and

$$((\omega\mathbf{n})f, f_\nu)=0,\qquad \nu=1, 2,\dots. \tag{21}$$

Taking the limit as $\nu\to\infty$ in this equation, we obtain

$$(\omega\mathbf{n}, f^2)=0. \tag{22}$$

Now let us take the scalar product of (20) and f and make use of relation (22):

$$(f, f)=(f, f)+\lambda(\omega\mathbf{n}, f^2)=(\hat{g}f, f)\leqslant g_0(f, f)\leqslant(f, f)=1.$$

This result and Theorem 4 imply that $g_0=1$ and $f(\omega)=1/2\sqrt{\pi}$ on Ω. Relation (20) now becomes

$$\frac{1}{2\sqrt{\pi}}(1+\lambda\omega\mathbf{n})=\frac{1}{2\sqrt{\pi}},\qquad \omega\in\Omega.$$

This means that $\lambda=0$. Moreover, the substitution $f(\omega)=1/2\sqrt{\pi}$ in (21) yields

$$(\omega\mathbf{n}, f_\nu)=0,\qquad \nu=1, 2,\dots. \tag{23}$$

Taking the scalar product of (19) and $\omega\mathbf{n}$, we obtain

$$(\omega\mathbf{n}, f_\nu)+k_\nu((\omega\mathbf{n})^2, f_\nu)=((\omega\mathbf{n}), \hat{g}f_\nu)=(\hat{g}(\omega\mathbf{n}), f_\nu),\ \nu=1, 2,\dots .$$

Equation (23) and the condition $k_\nu\neq\lambda=0$, $\nu=1, 2,\cdots$, imply that

$$((\omega\mathbf{n})^2, f_\nu)=\frac{1}{k_\nu}(\hat{g}(\omega\mathbf{n}), f_\nu),\qquad \nu=1, 2,\dots.$$

But $\omega\mathbf{n}=\sqrt{4\pi/3}\,Y^{(0)}_{\sigma 1}(\omega)$ (σ is an arbitrary orthonormal basis in R_3 with the third vector $\mathbf{n}$). Hence, by Theorem 2, $\hat{g}(\omega\mathbf{n})=g_1(\omega\mathbf{n})$, and, by (23), we have

$$((\omega\mathbf{n})^2, f_\nu)=0,\qquad \nu=1, 2,\dots.$$

Taking the limit as $\nu\to\infty$, we obtain

$$((\omega\mathbf{n})^2, f)=0.$$

But $f(\omega) \equiv \text{const} \neq 0$. This implies that

$$\|\omega \mathbf{n}\|^2 = ((\omega \mathbf{n})^2, 1) = 0$$

which is not true. This contradiction proves the theorem.

Does $\mathfrak{N}$ contain even one point? This question is easy to answer in the case $g_0 = 1$. In fact, for $\mathrm{k} = 0$ the characteristic equation becomes $f = \hat{g}f$. If $g_0 = 1$, then the function $f(\omega) \equiv 1$ satisfies this equation, which means that $0 \in \mathfrak{N}$. The problem becomes more complex in the general case.

We denote by $N(C(\Omega))$ the collection of all real functions continuous on Ω, $N(C(\Omega)) \subset C(\Omega)$. The space $N(C(\Omega))$ is a real Banach space with respect to the norm and the addition and multiplication operations induced from $C(\Omega)$. (The space $N(C(\Omega))$ introduced in this way is called the real kernel of $C(\Omega)$ [33].) Let K be the system of all functions nonnegative and continuous on the sphere Ω. Being a subset of $N(C(\Omega))$, the system K is a solid cone [34]. The function $x(\omega)$ defined on Ω is a point of K interior in the topology of $N(C(\Omega))$ if and only if $x(\omega)$ is continuous on Ω and $x(\omega) > 0$ for all $\omega \in \Omega$. We denote the collection of all interior points of K by $\operatorname{Int} K$.

The inclusion $\hat{U}(\mathrm{k})(L_2(\Omega)) \subset C(\Omega)$ enables us to define the operator $\hat{U}_C(\mathrm{k})$, $\mathrm{k} \in Z_0$, induced in $C(\Omega)$ in terms of the operator $\hat{U}(\mathrm{k})$: $\hat{U}_C(\mathrm{k})f = \hat{U}(\mathrm{k})$ for $f \in C(\Omega)$ and $\mathrm{k} \in Z_0$.

Lemma 7. *Let* $\mathrm{k} \in (-1, 1)$, *and let* $\hat{U}_C \equiv \hat{U}_C(\mathrm{k})$. *The operator* $\hat{U}_C$ *is a linear completely continuous operator acting from* $C(\Omega)$ *into* $C(\Omega)$. *The cone* K *is invariant with respect to* $\hat{U}_C$: $\hat{U}_C(K) \subset K$. *There exists a natural number* p *such that*

$$U_C^p x \in \operatorname{Int} K$$

for any function $x \in K$, $x \neq \vartheta$.

Proof. The linearity of U_C is self-evident, and its complete continuity and the inclusion $\hat{U}_C(C(\Omega)) \subset C(\Omega)$ follow from the relation $\hat{U}_C f(\omega) = (1/(1 + \mathrm{k}\omega\mathbf{n}))\hat{g}f(\omega)$, $f \in C(\Omega)$ and from Theorem 1. The kernel of U_C is the nonnegative function $g(\omega\omega')/(1 + \mathrm{k}\omega\mathbf{n})$. Hence $U_C x(\omega)$ is continuous and nonnegative on Ω for every $x \in K$. This means that $\hat{U}_C(K) \subset K$.

Now let us choose a natural number p such that $g_{(p)}(\mu) \geq g_0^p/5\pi$ for all $\mu \in [-1, 1]$. The existence of such a p follows from Theorem 4. If $x \in K$, then for every natural number n and for all $\omega \in \Omega$ we have

$$\hat{U}_C^n x(\omega) \geqslant (1 + |\mathrm{k}|)^{-n} \hat{g}^n x(\omega)$$

(by induction over n). In particular, for $x \in K$, $x \neq \vartheta$, for all $\omega \in \Omega$ (compare Lemma 4) we have

$$\hat{U}^p_C x(\omega) \geqslant (1+|k|)^{-p}\,\hat{g}_{(p)}x(\omega) \geqslant (1+|k|)^{-p}\,\frac{g_0^p}{5\pi}(x, 1) > 0.$$

This implies that $\hat{U}^p_C x \in \operatorname{Int} K$. The lemma is proved.

Lemma 7 enables us to use the theory of cones in Banach spaces [34] for studying the properties of the operator $\hat{U}_C(k)$. Relying on Theorem 6.3 of [34] and Lemma 7, we can assert that $\hat{U}_C(k)$, $k \in (-1, 1)$, has exactly one eigenfunction belonging to the cone K; the corresponding eigenvalue is simple and exceeds the modulus of any other spectral value of $\hat{U}_C(k)$. These properties of the operator $\hat{U}_C(k)$ can be readily extended to the operator $\hat{U}(k)\colon L_2(\Omega) \to L_2(\Omega)$. In fact, every eigenfunction of $\hat{U}(k)$ corresponding to a nonzero eigenvalue is continuous and is an eigenfunction of $\hat{U}_C(k)$ (Theorem 1). Making use of the complete continuity of $\hat{U}(k)$ and $\hat{U}_C(k)$, we infer from this, for example that $\sigma(\hat{U}(k))\setminus\{0\} = \sigma(\hat{U}_C(k))\setminus\{0\}$, $k \in (-1, 1)$. These simple considerations can be extended, and we now turn to the following result.

Theorem 6. *For* $k \in (-1, 1)$ *the operator* $\hat{U}(k)\colon L_2(\Omega) \to L_2(\Omega)$ *has the following properties.*

[i] *Let* $M(k) = \sup\{|\lambda| \mid \lambda \in \sigma(\hat{U}(k))\}$. *Then* $M(k) > 0$ *and* $M(k) \in \sigma(\hat{U}(k))$.

[ii] *If* $\lambda \in \sigma(\hat{U}(k))$, $\lambda \neq M(k)$, *then* $|\lambda| < M(k)$.

[iii] $M(k)$ *is a simple eigenvalue of* $\hat{U}(k)$.

[iv] *There exists exactly one function* $\Phi_k \in L_2(\Omega)$ *which satisfies the following two conditions:*

$$\hat{U}(k)\,\Phi_k = M(k)\,\Phi_k; \quad (\Phi_k, 1) = 4\pi.$$

[v] $\Phi_k \in \operatorname{Int} K$; *if* $f \in K\,(f \neq \vartheta)$ *and if* f *is an eigenfunction of the operator* $\hat{U}(k)$, *then there exists a number* $\alpha > 0$ *such that* $f = \alpha\Phi_k$.

The functions $M(k)$ and Φ_k are thus uniquely defined for all $k \in (-1, 1)$. If the number $k_1 \in (-1, 1)$ is a solution of the equation $M(k) = 1$, then clearly $k_1 \in \mathfrak{N}$.

Lemma 8. *The function* $M(k)$ *is continuous in the interval* $(-1, 1)$.

Proof. Let us consider the operator $\hat{V}(k)$, $k \in (-1, 1)$, defined by relation (17). $\hat{V}(k)$ acts from $L_2(\Omega)$ into $L_2(\Omega)$. Because $\hat{V}(k)$ is selfadjoint, it follows that

$$\|\hat{V}(k)\| = \sup\{|\lambda| \mid \lambda \in \sigma(\hat{V}(k))\} = \sup\{|(\hat{V}(k)f, f)| \mid f \in L_2(\Omega), \|f\| = 1\}.$$

$\hat{V}(k)$ is an integral operator with a nonnegative kernel. Hence

$$|(\hat{V}(k)f, f)| \leqslant (\hat{V}(k)|f|, |f|)$$

for every function $f \in L_2(\Omega)$. This means that

$$\sup\{(\hat{V}(k)f, f) \mid f \in L_2(\Omega), \|f\| = 1\}$$
$$= \sup\{|(\hat{V}(k)f, f)|\} = \|\hat{V}(k)\|, \quad k \in (-1,1).$$

Let $\lambda \in \sigma(\hat{V}(k))$. If $\lambda = 0$, then $|\lambda| < M(k)$. If $\lambda \neq 0$, then λ is an eigenvalue of $\hat{V}(k)$, and it is associated with an eigenfunction $f \in L_2(\Omega)$, $f \neq \vartheta$. We set $h(\omega) = f(\omega)/\sqrt{1 + k\omega n}$, $\omega \in \Omega$. Then $h \in L_2(\Omega)$, $h \neq \vartheta$ and

$$\hat{U}(k)h(\omega) = \frac{1}{\sqrt{1 + k\omega n}} \hat{V}(k) f(\omega) = \lambda h(\omega).$$

This means that $\lambda \in \sigma(\hat{U}(k))$ and $|\lambda| \leq M(k)$ (compare Theorem 6[i]). It is now clear that $\sup\{|\lambda| \mid \lambda \in \sigma(\hat{V}(k))\} \leq M(k)$.

Let $f(\omega) = \sqrt{1 + k\omega n}\, \Phi_k(\omega)$. Then $f \in L_2(\Omega)$, $f \neq \vartheta$, and $\hat{V}(k) f(\omega) = \sqrt{1 + k\omega n}\, \hat{U}(k)\, \Phi_k(\omega) = M(k) f(\omega)$. This means that $M(k) \in \sigma(\hat{V}(k))$ and $M(k) \leq \sup\{|\lambda| \mid \lambda \in \sigma(\hat{V}(k))\}$. Thus $M(k)$ can be expressed as

$$M(k) = \sup\{|\lambda| \mid \lambda \in \sigma(\hat{V}(k))\} = \|\hat{V}(k)\|$$
$$= \sup\{(\hat{V}(k)f, f) \mid f \in L_2(\Omega), \|f\| = 1\}, \quad k \in (-1,1).$$

Let us choose an $a \in (0, 1)$ and set $t = (1 - a)^{-1/2}$. To every function $f \in L_2(\Omega)$, $\|f\| = 1$, we assign an

$$s_f(k) = (\hat{V}(k)f, f), \ k \in (-1,1).$$

We note that

$$|(1 + k\omega n)^{-1/2}(1 + k\omega' n)^{-1/2} - (1 + k'\omega n)^{-1/2}(1 + k'\omega' n)^{-1/2}| \leqslant t^4 |k - k'|$$

for $\omega \in \Omega$, $\omega' \in \Omega$, $k \in [-a, a]$, $k' \in [-a, a]$. Let $f \in L_2(\Omega)$, $\|f\| = 1$, $k \in [-a, a]$, $k' \in [-a, a]$. We have

$$\|\hat{V}(k)f - \hat{V}(k')f\|^2$$
$$= \int_\Omega \Big| \int_\Omega [(1 + k\omega n)^{-1/2}(1 + k\omega' n)^{-1/2} - (1 + k'\omega n)^{-1/2}(1 + k'\omega' n)^{-1/2}]$$
$$\times g(\omega\omega') f(\omega')\, d\omega' \Big|^2 d\omega \leqslant 8\pi^2 t^8 |k - k'|^2 [g]^2.$$

Hence

$$|s_f(k) - s_f(k')| \leqslant \|\hat{V}(k)f - \hat{V}(k')f\| \leqslant T \cdot |k - k'|,$$
$$T = 2\sqrt{2}\,\pi t^4 [g].$$

Let us choose arbitrary k and k' from $[-a, a]$. Let $\epsilon > 0$. There exists an $f_0 \in L_2(\Omega)$, $\|f_0\| = 1$, such that $s_{f_0}(k) > M(k) - \epsilon$. We have

$$M(k) < s_{f_0}(k) + \varepsilon \leqslant s_{f_0}(k') + \varepsilon + T|k - k'| \leqslant M(k') + \varepsilon + T|k - k'|.$$

The arbitrariness of ϵ implies that $M(k) - M(k') \leq T|k - k'|$. Transposing k and k' and repeating the above argument, we find that

$$|M(k) - M(k')| \leqslant T \cdot |k - k'|, \quad k \in [-a, a], \quad k' \in [-a, a].$$

The lemma is proved.

We note now that $\hat{U}(k)$ is a holomorphic function [35] of the variable $k \in Z_0$ with values in the Banach space of linear bounded operators acting in $L_2(\Omega)$ (compare (16)). This enables us to expand $\hat{U}(k)$ in a series in powers of $k - k_0$, $k_0 \in Z_0$, which converges in the operator norm for sufficiently small $k - k_0$. If $k_0 \in (-1, 1)$, then $\hat{U}(k) = \Sigma_{\nu=0}^{\infty} (k - k_0)^{\nu} \hat{U}_{\nu}$ for $|k - k_0| < 1 - |k_0|$. The operator $\hat{U}_{\nu}$ is a linear bounded operator acting in $L_2(\Omega)$; and $\hat{U}_{\nu} = (-1)^{\nu}((\omega\mathbf{n})^{\nu}/(1 + k_0\omega\mathbf{n})^{\nu+1})\,\hat{g}$, $\|\hat{U}_{\nu}\| \leq 2\pi\sqrt{2}[g]\,(1 - |k_0|)^{-(\nu+1)}$ for $\nu = 0, 1, 2, \cdots$. The point $\lambda = M(k_0)$ is an isolated point of the spectrum $\sigma(\hat{U}_0)$ ($\hat{U}_0 = \hat{U}(k_0)$); and $M(k_0)$ is a simple eigenvalue of the operator $\hat{U}_0$.

The above properties of the eigenvalue $M(k_0)$ and of the operators $\hat{U}_{\nu}$, $\nu \geq 0$, are sufficient to enable us to apply the theory of analytic perturbations of linear operators [36] to $\hat{U}(k)$. According to this theory there exist a number $h \in (0, 1 - |k_0|)$ and a circle γ on the complex λ-plane with its center at the point $M(k_0)$ such that

a) For $k \in S(k_0, h) \equiv \{k' \,|\, |k' - k_0| < h\}$ the circle γ lies in the resolvent set of the operator $\hat{U}(k)$.

b) For $k \in S(k_0, h)$ the circle γ contains exactly one point of the spectrum $\sigma(\hat{U}(k))$. This point $\lambda(k)$ is a simple eigenvalue of the operator $\hat{U}(k)$. Being a function of k, the point $\lambda(k)$ is holomorphic in $S(k_0, h)$.

c) A function $e(k)$ with values in $L_2(\Omega)$ is defined in the circle $S(k_0, h)$. The function $e(k)$ is holomorphic in $S(k_0, h)$ and $\|e(k)\| > 0$, $\hat{U}(k)\,e(k) = \lambda(k)\,e(k)$ for all $k \in S(k_0, h)$.

Let us denote the radius of the circle γ by r and let $h_1 \leq h$ be a positive number such that $|M(k) - M(k_0)| < r$ for $k_0 - h_1 < k < k_0 + h_1$. The existence of such an h_1 follows from Lemma 8. Since both $M(k)$ and $\lambda(k)$ lie inside the circle γ for $k_0 - h_1 < k < k_0 + h_1$, it follows by property b) that $\lambda(k) = M(k)$ for every real k, $|k - k_0| < h_1$. For the same k we have $\|e(k)\| > 0$ and $\hat{U}(k)\,e(k) = M(k)\,e(k)$. From Theorem 6[iii],[iv] we find that

$$e(k) = \frac{(e(k), 1)}{4\pi} \Phi_k, \quad |(e(k), 1)| > 0$$

for $k_0 < h_1 < k < k_0 + h_1$.

The numerical function $(e(k), 1)$ is holomorphic in $S(k_0, h)$. Reducing h_1 if need be, we can assume that $\lambda(k) \neq 0$ and $(e(k), 1) \neq 0$ in the circle $S(k_0, h_1)$.

Let $e_0(k) = (4\pi/(e(k), 1))\, e(k)$. It is clear that $e_0(k)$ and $\lambda(k)$ are holomorphic in $S(k_0, h_1)$, and that $e_0(k) = \Phi_k$ and $\lambda(k) = M(k)$ on the portion of the real axis lying inside $S(k_0, h_1)$.

Thus Φ_k and $M(k)$ have been extended analytically onto some neighborhood of the point k_0. Taking an arbitrary $k_0 \in (-1, 1)$, we arrive at the following theorem.

Theorem 7. *There exist an open set* G_0, $(-1, 1) \subset G_0 \subset Z_0$, *and functions* $\lambda(k)$ *and* $e_0(k)$ *which are defined for* $k \in G_0$ *and have the following properties.*

[i] $\lambda(k)$ *is a numerical function;* $e_0(k)$ *is a function with values in* $L_2(\Omega)$; *and* $\lambda(k)$ *and* $e_0(k)$ *are holomorphic in* G_0.

[ii] $\lambda(k) \neq 0$, $\|e_0(k)\| > 0$, *and* $\hat{U}(k)\, e_0(k) = \lambda(k)\, e_0(k)$ *for* $k \in G_0$.

[iii] $\lambda(k) = M(k)$ *and* $e_0(k) = \Phi_k$ *for* $k \in (-1, 1)$.

It is clear that the conditions of Theorem 7 define the functions $\lambda(k)$ and $e_0(k)$ uniquely. For $k_1 \in G_0$ and small $k - k_1$ the function $e_0(k)$ can be expanded in a series in powers of $k - k_1$ which converges in the metric of $L_2(\Omega)$. This result can be strengthened: the Taylor series for $e_0(k)$ converges uniformly, and its coefficients are continuous functions of the variable $\omega \in \Omega$.

Theorem 8. *Let* G_0 *and* $e_0(k)$ *be the set and the function defined in Theorem* 8. *Let*

$$e_0^{[q]}(k) = \frac{d^q}{dk^q} e_0(k), \quad q = 0, 1, 2, \ldots$$

for $k \in G_0$. *Then* $e_0^{[q]}(k) \in C(\Omega)$ *for all* $k \in G_0$, $q = 0, 1, 2, \cdots$. *To every* $k_1 \in G_0$ *we can assign an* $r = r(k_1) > 0$ *such that*

$$\| e_0^{[q]}(k_1) \|_{C(\Omega)} \leqslant C_0 q!\, r^{-q}, \quad q = 0, 1, 2. \ldots, \quad C_0 = \text{const} > 0,$$

and

$$\lim_{N=\infty} \left\| e_0(k) - \sum_{q=0}^{N} \frac{1}{q!} (k - k_1)^q e_0^{[q]}(k_1) \right\|_{C(\Omega)} = 0$$

uniformly for $k \in S(k_1, r')$, $r' \in (0, r)$.

Proof. Let $k_1 \in G_0$. Because $e_0(k)$ and $\lambda(k)$ are analytic in G_0, there exist numbers $r_1 > 0$ and $C_1 > 0$ such that

$$e_0^{[q]}(k_1) \in L_2(\Omega), \quad \| e_0^{[q]}(k_1) \| \leqslant \frac{C_1 q!}{r_1^q}, \quad | \lambda^{[q]}(k_1) | \leqslant \frac{C_1 q!}{r_1^q} \tag{24}$$

($\lambda^{[q]}(k) = d^q \lambda(k)/dk^q$) for all $q = 0, 1, 2, \cdots$. Moreover,

$$\lambda(k)(1 + k\omega n)\, e_0(k) = \hat{g} e_0(k) \tag{25}$$

everywhere in G_0. It is not difficult to justify the differentiability of both sides of this equation according to the usual rules. For $q \geq 1$ we have

$$\lambda(\mathrm{k})(1+\mathrm{k}\omega\mathrm{n})e_0^{[q]}(\mathrm{k}) + \sum_{s=1}^{q}\binom{q}{s}[\lambda^{[s]}(\mathrm{k})(1+\mathrm{k}\omega\mathrm{n}) + s\lambda^{[s-1]}(\mathrm{k})(\omega\mathrm{n})]e_0^{[q-s]}(\mathrm{k}) = \hat{g}e_0^{[q]}(\mathrm{k}).$$

Since $\lambda(\mathrm{k}) \neq 0$, $\mathrm{k} \in G_0$, we have

$$e_0^{[q]}(\mathrm{k}) = \frac{1}{\lambda(\mathrm{k})(1+\mathrm{k}\omega\mathrm{n})}\Big\{\hat{g}e_0^{[q]}(\mathrm{k}) - \sum_{s=1}^{q}\binom{q}{s}[\lambda^{[s]}(\mathrm{k})(1+\mathrm{k}\omega\mathrm{n}) + s\lambda^{[s-1]}(\mathrm{k})(\omega\mathrm{n})]e_0^{[q-s]}(\mathrm{k})\Big\}. \tag{26}$$

The inclusion $e_0^{[0]}(\mathrm{k}) = e_0(\mathrm{k}) \in C(\Omega)$, $\mathrm{k} \in G_0$ follows from (25) and Theorem 1. By induction over q and with the aid of (26) and Theorem 1 we can establish that $e_0^{[q]}(\mathrm{k}) \in C(\Omega)$ for all $\mathrm{k} \in G_0$ and $q \geq 1$.

Let us set $N_q = \|e_0^{[q]}(\mathrm{k}_1)\|_{C(\Omega)}$, $q = 0, 1, 2, \cdots$, and let $j_+ = \sup_{\omega\in\Omega}|1 + \mathrm{k}_1\omega\mathrm{n}|$, $j_- = \inf_{\omega\in\Omega}|1 + \mathrm{k}_1\omega\mathrm{n}|$. Then $j_+ \geq j_- > 0$, and for $q \geq 1$ we find (from (26) and (24)) that

$$N_q \leqslant \frac{1}{|\lambda(\mathrm{k}_1)|\,j_-}\Big\{\sqrt{2\pi}\,[g]\frac{C_1 q!}{r_1^q} + \sum_{s=1}^{q}\binom{q}{s}\Big[j_+\frac{C_1 s!}{r_1^s} + s\frac{C_1(s-1)!}{r_1^{s-1}}\Big]N_{q-s}\Big\}$$
$$\leqslant C\Big[\frac{q!}{r_1^q} + \sum_{s=1}^{q}\binom{q}{s}\frac{s!}{r_1^s}N_{q-s}\Big], \quad q = 1, 2, \ldots,$$

where

$$C = \frac{C_1}{|\lambda(k_1)|\,j_-}\max\{\sqrt{2\pi}\,[g],\ r_1 + j_+\}.$$

Let $\mu_q = r_1^q N_q/q!$, $q = 0, 1, 2, \cdots$. This inequality implies that $\mu_q \leq C(1 + \Sigma_{s=0}^{q-1}\mu_s)$, $q = 1, 2, \cdots$. Hence $\mu_q \leq C(1 + C)^{q-1}(1 + \mu_0)$, $q = 1, 2, \cdots$, so that $N_q \leq C_0 q!\, r^{-q}$, $q = 1, 2, \cdots$, $C_0 = C(1 + N_0)/(1 + C)$ and $r = r_1/(1 + C)$.

Let r' be any number from the interval $(0, r)$. The series

$$\sum_{q=0}^{\infty}\frac{1}{q!}(\mathrm{k}-\mathrm{k}_1)^q e_0^{[q]}(\mathrm{k}_1) = f(\mathrm{k}) \in C(\Omega)$$

then converges in the metric of $C(\Omega)$ uniformly for all $\mathrm{k} \in S(\mathrm{k}_1, r')$. For a fixed $\mathrm{k} \in S(\mathrm{k}_1, r')$ we have

$$\|e_0(\mathrm{k}) - f(\mathrm{k})\| \leqslant \Big\|e_0(\mathrm{k}) - \sum_{q=0}^{N}\frac{1}{q!}(\mathrm{k}-\mathrm{k}_1)^q e_0^{[q]}(\mathrm{k}_1)\Big\| + 2\sqrt{\pi}\Big\|\sum_{a=0}^{N}\frac{1}{q!}(\mathrm{k}-\mathrm{k}_1)^q e_0^{[q]}(\mathrm{k}_1) - f(\mathrm{k})\Big\|_{C(\Omega)}.$$

The right side vanishes as $N \to \infty$ (since $|k - k_1| < r' < r \le r_1$). This implies that $f(k) = e_0(k)$, and the theorem is proved.

Corollary. $\lim_{k=k_1} \|e_0(k) - e_0(k_1)\|_{C(\Omega)} = 0$ *for every* $k_1 \in G_0$. *In particular,*

$$\lim_{k=k_1} \Phi_k(\omega) = \Phi_{k_1}(\omega)$$

uniformly in all $\omega \in \Omega$ *for* $k_1 \in (-1, 1)$.

In fact, for $|k - k_1| < r$ we have

$$\|e_0(k) - e_0(k_1)\|_{C(\Omega)} \leqslant \sum_{q=1}^{\infty} \frac{1}{q!} |k - k_1|^q \|e_0^{[q]}(k_1)\|_{C(\Omega)} \leqslant C_0 \frac{|k - k_1|}{r - |k - k_1|}.$$

The analyticity of $M(k)$ and Φ_k in $(-1, 1)$ established in Theorem 7 enables us to prove the solvability of the equation $M(k) = 1$ and at the same time the nonemptiness of the class $\mathfrak{N}$.

Theorem 9. $M(0) = g_0$; $M(-k) = M(k)$ *for* $k \in (-1, 1)$; $M'(k) \equiv dM(k)/dk > 0$ *for* $k \in (0, 1)$; $M'(0) = 0$; *and* $\lim_{k=1-0} M(k) = \infty$.

Proof. By definition (see Theorem 6), $M(0) = \sup\{|\lambda| \mid \lambda \in \sigma(\hat{g})\}$. But $\sigma(\hat{g}) = \{g_n \mid n = 0, 1, 2, \cdots\} \cup \{0\}$ (Theorem 2[ii]), whence it follows that $M(0) = g_0$ (compare (14)).

Since $M(k) > 0$ and $M(k) \in \sigma(\hat{U}(k))$, we can write

$$M(k) = \sup\{|\lambda| \mid \lambda \in \sigma(\hat{U}(k)) \setminus \{0\}\}, \quad k \in (-1, 1).$$

This and Lemma 6[ii] imply the evenness of $M(k)$. $M(-k) = M(k)$, $k \in (-1, 1)$. The equation $M'(0) = 0$ follows from the evenness and analyticity of $M(k)$.

We can now prove that the derivative $M'(k)$ is positive for $k \in (0, 1)$. Let us rewrite (25) and (26) for $q = 1$ and $k \in (-1, 1)$,

$$M(k)(1 + k\omega n)\Phi_k = \hat{g}\Phi_k. \tag{27}$$

$$M'(k)(1 + k\omega n)\Phi_k + M(k)\omega n\Phi_k + M(k)(1 + k\omega n)\Phi_k' = \hat{g}\Phi_k'. \tag{28}$$

Here $\Phi_k' = d\Phi_k/dk$, $k \in (-1, 1)$. From the fact that $\Phi_k(\omega)$ is a real function and that $\Phi_k'(\omega) = \text{l.i.m.}_{n=\infty} [n(\Phi_{k+1/n}(\omega) - \Phi_k(\omega))]$, we infer that $\Phi_k'(\omega)$ is a real function.

We note that $(\hat{g}\Phi_k', \Phi_k) = (\Phi_k', \hat{g}\Phi_k) = M(k)(\Phi_k', (1 + k\omega n)\Phi_k) = M(k)((1 + k\omega n)\Phi_k', \Phi_k)$. Hence, taking the scalar product of the right side of (28) and Φ_k', we obtain

$$M'(k)[\|\Phi_k\|^2 + k(\omega n, \Phi_k^2)] + M(k)(\omega n, \Phi_k^2) = 0. \tag{29}$$

We note that $|(\omega n, \Phi_k^2)| \le \|\Phi_k\|^2$. Hence $|k(\omega n, \Phi_k^2)| < \|\Phi_k\|^2$, $k \in (-1, 1)$. This enables us to solve (29) for $M'(k)$:

$$M'(\mathrm{k}) = -M(\mathrm{k})\frac{(\omega\mathbf{n},\ \Phi_{\mathrm{k}}^2)}{\|\Phi_{\mathrm{k}}\|^2 + \mathrm{k}\,(\omega\mathbf{n},\ \Phi_{\mathrm{k}}^2)},\quad \mathrm{k}\in(-1,\,1). \tag{30}$$

Now let us investigate the behavior of Φ_{k} for small k. According to Theorem 7 there exists an $h > 0$ such that

$$\Phi_{\mathrm{k}} = \Phi_0 + \mathrm{k}\Phi_0' + \mathrm{k}^2\alpha(\mathrm{k}),\qquad \sup_{-h\leqslant \mathrm{k}\leqslant h}\|\alpha(\mathrm{k})\| = \alpha^+ < \infty \tag{31}$$

for $-h \le \mathrm{k} \le h$. We set $\mathrm{k} = 0$ in (27) and recall that $M(0) = g_0$ (see above):

$$g_0\Phi_0 = \hat{g}\Phi_0.$$

According to Theorem 4, $\Phi_0(\omega) \equiv C = \text{const},\ \omega \in \Omega$. From the normalization condition $(\Phi_0,\ 1) = 4\pi$ (Theorem $6^{[iv]}$) we find that

$$\Phi_0(\omega) \equiv 1 \qquad \omega\in\Omega. \tag{32}$$

Equation (28) for $k = 0$ now becomes

$$g_0\omega\mathbf{n} + g_0\Phi_0'(\omega) = \hat{g}\Phi_0'(\omega).$$

Let us take the scalar product of the right side of this equation and $\omega\mathbf{n}$:

$$g_0(\omega\mathbf{n},\ \omega\mathbf{n}) + g_0(\Phi_0',\ \omega\mathbf{n}) = (\hat{g}\Phi_0',\ \omega\mathbf{n}) = (\Phi_0',\ \hat{g}(\omega\mathbf{n})) = g_1(\Phi_0',\ \omega\mathbf{n}),$$

$$(g_0 - g_1)(\Phi_0',\ \omega\mathbf{n}) = -g_0(\omega\mathbf{n},\ \omega\mathbf{n}).$$

It is clear that $(\omega\mathbf{n},\ \omega\mathbf{n}) > 0$. Moreover, $g_0 > 0$ and $g_0 > g_1$ (14). Hence

$$(\Phi_0',\ \omega\mathbf{n}) < 0. \tag{33}$$

For $\mathrm{k} \neq 0$, $-h \le \mathrm{k} \le h$, we have

$$\frac{(\omega\mathbf{n},\ \Phi_{\mathrm{k}}^2)}{\mathrm{k}} = \frac{((\omega\mathbf{n})\,\Phi_{\mathrm{k}},\ \Phi_{\mathrm{k}})}{k}$$

$$= \frac{1}{\mathrm{k}}((\omega\mathbf{n})\,\Phi_0 + \mathrm{k}\,(\omega\mathbf{n})\,\Phi_0' + \mathrm{k}^2(\omega\mathbf{n})\,\alpha(\mathrm{k}),\ \Phi_0 + \mathrm{k}\Phi_0' + \mathrm{k}^2\alpha(\mathrm{k}))$$

$$= \frac{1}{\mathrm{k}}(\omega\mathbf{n},\ 1) + 2\,(\Phi_0',\ \omega\mathbf{n}) + \mathrm{k}D(\mathrm{k}),$$

$$D(\mathrm{k}) = ((\omega\mathbf{n})\,\Phi_0',\ \Phi_0') + 2\,(\omega\mathbf{n},\ \alpha(\mathrm{k})) + 2\mathrm{k}\,((\omega\mathbf{n})\,\Phi_0',\ \alpha(\mathrm{k})) + \mathrm{k}^2((\omega\mathbf{n})\,\alpha(\mathrm{k}),\ \alpha(\mathrm{k})).$$

But $(\omega\mathbf{n},\ 1) = 2\pi\int_{-1}^{1}\mu\,d\mu = 0$, and for $-h \le \mathrm{k} \le h$ we obtain

$$|D(\mathrm{k})| \leqslant \|\Phi_0'\|^2 + 4\sqrt{\frac{\pi}{3}}\,\alpha^+ + 2h\alpha^+\|\Phi_0'\| + h^2(\alpha^+)^2 = \text{const} < \infty.$$

Thus

$$\lim_{\mathrm{k}=0}\frac{(\omega\mathbf{n},\ \Phi_{\mathrm{k}}^2)}{\mathrm{k}} = 2\,(\Phi_0',\ \omega\mathbf{n}). \tag{34}$$

Hence $\lim_{k=\infty}[k(\omega\mathbf{n}, \Phi_k^2)] = 0$. At the same time, (31) and (32) imply that $\lim_{k=0}\|\Phi_k\|^2 = \|\Phi_0\|^2 = 4\pi$. Making use of these limit relations and formula (30), we find that

$$\lim_{k=0}\frac{M'(k)}{k} = -\frac{g_0}{2\pi}(\Phi_0', \omega\mathbf{n}) > 0$$

(compare (33)). This means that there exists a $k_1 \in (0, 1)$ such that $M'(k) > 0$ for $k \in (0, k_1)$.

Let us suppose that the inequality $M'(k) > 0$ is fulfilled only for some $k \in (0, 1)$. There then exists a number $k_2 \in [k_1, 1)$ for which $M'(k_2) = 0$, but $M'(k) > 0$ for $k \in (0, k_2)$. It is clear that

$$M(k_2) > M(0) = g_0. \qquad (*)$$

From (28) we find that $(\omega\mathbf{n}, \Phi_{k_2}^2) = 0$ for $k = k_2$. We take the scalar product of (27) and Φ_k and set $k = k_2$ in the resulting expression:

$$(\hat{g}\Phi_{k_2}, \Phi_{k_2}) = M(k_2)\|\Phi_{k_2}\|^2 + k_2 M(k_2)(\omega\mathbf{n}, \Phi_{k_2}^2) = M(k_2)\|\Phi_{k_2}\|^2.$$

But $(\hat{g}\Phi_{k_2}, \Phi_{k_2}) \leq \|\hat{g}\|\,\|\Phi_{k_2}\|^2 = g_0\|\Phi_{k_2}\|^2$, which means that

$$M(k_2) \leq g_0.$$

But this contradicts inequality (*), which implies that the inequality $M'(k) > 0$ is fulfilled at all points $k \in (0, 1)$.

It remains for us to verify the limit relation $\lim_{k=1-0} M(k) = \infty$. In proving Lemma 7 we isolated a natural number p such that $g_{(p)}(\mu) \geq g_0^p/5\pi$ for all $\mu \in [-1, 1]$. We shall now show that if $f \in L_2(\Omega)$ and $f(\omega) \geq 0$ on Ω, then

$$\hat{V}^p(k)f(\omega) \geqslant \frac{g_0^p}{5\pi\cdot 2^{p-1}}\frac{1}{\sqrt{1+k\omega\mathbf{n}}}\int_\Omega \frac{f(\omega')\,d\omega'}{\sqrt{1+k\omega'\mathbf{n}}}, \quad k\in(-1,1), \quad \omega\in\Omega. \quad (35)$$

In fact, $\hat{V}(k)f(\omega) = (1/\sqrt{1+k\omega\mathbf{n}})\,\hat{g}(f(\omega)/\sqrt{1+k\omega\mathbf{n}})$ (compare (16) and (17)). If

$$\hat{V}^s f(\omega) \geqslant 2^{-(s-1)}\frac{1}{\sqrt{1+k\omega\mathbf{n}}}\hat{g}^s\frac{f(\omega)}{\sqrt{1+k\omega\mathbf{n}}}, \qquad (36)$$

for some $s \geq 1$ and for a nonnegative $f \in L_2(\Omega)$, then

$$\hat{V}^{s+1}f(\omega) \geqslant 2^{-(s-1)}\frac{1}{\sqrt{1+k\omega\mathbf{n}}}\hat{g}\frac{1}{\sqrt{1+k\omega\mathbf{n}}}\left(\frac{1}{\sqrt{1+k\omega\mathbf{n}}}\hat{g}^s\frac{f(\omega)}{\sqrt{1+k\omega\mathbf{n}}}\right)$$
$$\geqslant 2^{-s}\frac{1}{\sqrt{1+k\omega\mathbf{n}}}\hat{g}^{s+1}\frac{f(\omega)}{\sqrt{1+k\omega\mathbf{n}}}.$$

Thus formula (36) is valid for all nonnegative functions $f \in L_2(\Omega)$ and for all natural numbers s. The operator $\hat{g}^s$ is an integral operator with the kernel

$g_{(s)}(\omega\omega')$ (Lemma 4). Hence (35) follows from (36) for $s = p$.

Since $M(k)$ is even and $\hat{V}(k)$ is selfadjoint, we can write (compare the proof of Lemma 8)

$$M^p(k) \geqslant \frac{g_0^p}{5\pi\cdot 2^{p-1}} \int_\Omega \frac{d\omega}{1 - k\omega n} = \frac{g_0^p}{5\cdot 2^{p-2}} \cdot \frac{1}{k} \ln \frac{1+k}{1-k}, \quad k \in (-1,1).$$

This means that

$$M^p(k) \geqslant \frac{(\hat{V}^p(-k)f, f)}{(f, f)}$$

for every function $f \in L_2(\Omega)$, $f \neq \vartheta$. This and (35) imply that for a fixed $k \in (-1, 1)$ we have

$$M^p(k) \geqslant \frac{g_0^p}{5\pi\cdot 2^{p-1}} \cdot \frac{\left(\int_\Omega \frac{f(\omega)}{\sqrt{1-k\omega n}}\, d\omega\right)^2}{\int_\Omega f^2(\omega)\, d\omega} \tag{37}$$

for every nonnegative function $f \in L_2(\Omega)$, $f \neq \vartheta$. Clearly the function $f(\omega) = (1 - k\omega n)^{-1/2}$ satisfies the conditions of applicability of formula (37), so that

$$M^p(k) = M^p(-k) = \|\hat{V}(-k)\|^p = \|\hat{V}^p(-k)\|.$$

This means that $M(k) \to \infty$ as $k \to 1 - 0$. The theorem is proved.

Corollary. *There exists exactly one* $\lambda_0 \in [0, 1)$ *such that* $M(\lambda_0) = 1$. *If* $g_0 = 1$, *then* $\lambda_0 = 0$; *if* $g_0 < 1$, *then* $\lambda_0 > 0$.

We can now supplement Theorem 5.

Theorem 10. $\pm\lambda_0 \in \mathfrak{N}$, *and* $\mathfrak{N} \cap (-\lambda_0, \lambda_0) = \emptyset$.

The set $\mathfrak{N}$ is therefore nonempty and lies entirely in the difference between two intervals on the real axis:

$$\mathfrak{N} \subset (-1,1) \setminus (-\lambda_0, \lambda_0).$$

Proof. We have $\pm\lambda_0 \in (-1, 1) \subset Z_0$, and $1 = M(\pm\lambda_0) \in \sigma(\hat{U}(\pm\lambda_0))$. This means that $\pm\lambda_0 \in \mathfrak{N}$. Let us suppose that there exists a $k \in \mathfrak{N} \cap (-\lambda_0, \lambda_0)$. According to Theorem 5, we can assume without limiting generality that $k \geq 0$. The fact that $1 \in \sigma(U(k))$ implies that

$$M(k) = \sup\{|\lambda| \mid \lambda \in \sigma(\hat{U}(k))\} \geqslant 1 = M(\lambda_0).$$

At the same time $0 \leq k < \lambda_0$, and, by Theorem 9, we have $M(k) < M(\lambda_0)$. The resulting contradiction proves the theorem.

Theorem 11. $\Phi_0(\omega) \equiv 1$, $\omega \in \Omega$. *For* $k \in (-1, 1)$ *and* $\omega \in \Omega$ *we have*

$$\Phi_k(-\omega) = \Phi_{-k}(\omega).$$

Proof. The first statement has already been proved (compare (32)). Let $\mathrm{k} \in (-1, 1)$. We set $\psi(\omega) = \Phi_{\mathrm{k}}(-\omega)$. Then $\psi \in C(\Omega)$, $(\psi, 1) = \int_\Omega \psi(\omega)d\omega = 4\pi$ (Lemma 1).

Moreover (compare Lemma 6[i]),

$$\hat{U}(-\mathrm{k})\psi(\omega) = \hat{U}(\mathrm{k})\Phi_{\mathrm{k}}(-\omega) = M(\mathrm{k})\Phi_{\mathrm{k}}(-\omega) = M(-\mathrm{k})\psi(\omega).$$

By Theorem 6[iv] these conditions imply that $\psi = \Phi_{-\mathrm{k}}$.

Theorem 12. *Let* $\mathrm{k} \in \mathfrak{R}$ *and let* $f(\omega)$ *be a real solution of the characteristic equation* $(1 + \mathrm{k}\omega\mathbf{n}) f(\omega) = \hat{g}f(\omega)$, $f \in L_2(\Omega)$, $f \neq \vartheta$. *Then* $f \in C(\Omega)$ *and*

[i] *If* $f(\omega) \geq 0$ *on* Ω, *then either* $\mathrm{k} = \lambda_0$ *and* $f = \alpha_1\Phi_{\lambda_0}$ $(\alpha_1 = \mathrm{const} > 0)$, *or* $\mathrm{k} = -\lambda_0$ *and* $f = \alpha_2\Phi_{-\lambda_0}$ $(\alpha_2 = \mathrm{const} > 0)$.

[ii] *If* $|\mathrm{k}| \neq \lambda_0$, *then* $f(\omega)$ *assumes both positive and negative values on* Ω.

Proof. The inclusion $f \in C(\Omega)$ follows from Theorem 1. Let us accept assumptions [i]. Then f belongs to the cone K, $f \neq \vartheta$. At the same time $\hat{U}(\mathrm{k})f = f$. Theorem 6[v] implies that $f = \alpha\Phi_{\mathrm{k}}$, where α is a positive number. We have $M(\mathrm{k})\Phi_{\mathrm{k}} = \hat{U}(\mathrm{k})\Phi_{\mathrm{k}} = \alpha^{-1}\hat{U}(\mathrm{k})f = \alpha^{-1}f = \Phi_{\mathrm{k}}$, which implies that $M(\mathrm{k}) = 1$. By the corollary of Theorem 9, this means that either $\mathrm{k} = \lambda_0$ or $\mathrm{k} = -\lambda_0$. The formula $f = \alpha\Phi_{\mathrm{k}}$ becomes $f = \alpha\Phi_{\lambda_0}$ in the former case and $f = \alpha\Phi_{-\lambda_0}$ in the latter. [i] has been proved. Statement [ii] is a self-evident consequence of [i].

§5. Separation of variables in the solutions of the characteristic equation

As in the previous section, we shall assume that the indicatrix g and the unit vector $\mathbf{n}$ appearing in characteristic equation (15) are fixed.

We denote by D, D_1, and D_2 segments lying in the spaces R_2, R_3, and R_4, respectively,

$$D = [-1,1] \times [-1,1];$$
$$D_1 = [-1,1] \times \Delta \equiv [-1,1] \times [-1,1] \times [0, 2\pi];$$
$$D_2 = \Delta^2 \equiv [-1,1] \times [0, 2\pi] \times [-1,1] \times [0, 2\pi].$$

Let $\gamma(\mu, \phi, \mu', \phi') = \mu\mu' + \sqrt{1-\mu^2}\sqrt{1-\mu'^2}\cos(\phi - \phi')$ and let $\gamma_0(\mu, \mu', \phi) = \gamma(\mu, \phi, \mu', 0)$. It is clear that γ (γ_0, respectively) is a continuous mapping of the segment D_2 (of the segment D_1, respectively) onto the segment $[-1, 1]$. If G is the open part of the segment $[-1, 1]$, then $H = \gamma^{-1}(G)$ is the open part of D_2. Let $a(\mu)$ be the characteristic function of G:

$$a(\mu) = \begin{cases} 1 & \text{for } \mu \in G, \\ 0 & \text{for } \mu \in [-1,1] \setminus G, \end{cases}$$

and let $A(\mu, \phi, \mu', \phi') = a(\gamma(\mu, \phi, \mu', \phi'))$. Then $A(\mu, \phi, \mu', \phi')$ is the characteristic function of H. Making use of the Fubini theorem, we can write

$$\operatorname{mes} H = \int_{\Delta} d\mu\, d\varphi \int_{\Delta} A(\mu, \varphi, \mu', \varphi')\, d\mu'\, d\varphi'.$$

Let us choose an orthonormal basis σ of the space R_3. Let $\omega_0 = \psi_\sigma(\mu, \phi)$. Then $A(\mu, \phi, \mu', \phi') = a(\omega_0 \psi_0(\mu', \phi'))$. By virtue of Lemma 1 we have

$$\int_{\Delta} A(\mu, \varphi, \mu', \varphi')\, d\mu'\, d\varphi' = \int_{\Omega} a(\omega_0\omega)\, d\omega = 2\pi \int_{-1}^{1} a(\mu)\, d\mu = 2\pi \operatorname{mes} G.$$

Thus, $\operatorname{mes} H = 8\pi^2 \operatorname{mes} G$. This means that if $E \subset [-1, 1]$, $\mathfrak{E} = \gamma^{-1}(E)$, and $\operatorname{mes} E = 0$, then $\operatorname{mes} \mathfrak{E} = 0$. Let $\{h_n(\mu)\}$ be a sequence of functions continuous on $[-1, 1]$ which converges to $g(\mu)$ almost everywhere. By the above considerations, $h_n(\gamma(\mu, \phi, \mu', \phi')) \to g(\gamma(\mu, \phi, \mu', \phi'))$ as $n \to \infty$ almost everywhere on D_2. The functions $h_n(\gamma(\mu, \phi, \mu', \phi'))$ are continuous on D_2. This means that $g(\gamma(\mu, \phi, \mu', \phi'))$ is measurable on D_2. Moreover,

$$\int_{D_2} |g(\gamma(\mu, \varphi, \mu', \varphi'))|^2 d\mu\, d\varphi\, d\mu'\, d\varphi' = \int_{\Delta} d\mu\, d\varphi \int_{\Omega} |g(\omega_0\omega)|^2\, d\omega = 8\pi^2 [g]^2 < \infty.$$

We can apply the same analysis to the functions $g(\gamma_0(\mu, \mu', \phi))$ on D_1. This yields the following lemma.

Lemma 9. *The functions $g(\gamma(\mu, \phi, \mu', \phi'))$ and $g(\gamma_0(\mu, \mu', \phi))$ are quadratically summable on the segments D_2 and D_1, respectively.*

Let s be an integer. We are now concerned with the convergence of the series

$$\sum_{n=|s|}^{\infty} \frac{2n+1}{2} \frac{(n-|s|)!}{(n+|s|)!} g_n P_n^{(s)}(\mu) P_n^{(s)}(\mu'). \tag{38}$$

Here the g_n are the coefficients of the expansion of the indicatrix in a series in Legendre polynomials (Theorem 2).

Lemma 10. *Series* (38) *converges in the metric of $L_2(D)$. If we consider it as a series in functions of the variables $\mu' \in [-1, 1]$ which depend on μ as on a parameter, then series* (38) *converges in the metric of $L_2(-1, 1)$ uniformly with respect to all $\mu \in [-1, 1]$.*

Proof. Let

$$S_{N_1 N_2}(\mu, \mu') = \sum_{n=N_1}^{N_2} \frac{2n+1}{2} \frac{(n-|s|)!}{(n+|s|)!} g_n P_n^{(s)}(\mu) P_n^{(s)}(\mu'), \quad N_2 > N_1 \geqslant |s|.$$

It is not difficult to show that $\int_D |\tilde{S}_{N_1 N_2}(\mu, \mu')|^2 d\mu\, d\mu' = \sum_{n=N_1}^{N_2} g_n^2$. The convergence of (38) in the metric of $L_2(D)$ now follows from the fact that

$\sum_{n=0}^{\infty} n g_n^2 < \infty$ (compare Theorem 2).

Let us choose a $\mu \in [-1, 1]$ and set $\sigma_{N_1 N_2}(\mu') = S_{N_1 N_2}(\mu, \mu')$. Then

$$[\sigma_{N_1 N_2}]^2 = \sum_{n=N_1}^{N_2} \frac{2n+1}{2} \frac{(n-|s|)!}{(n+|s|)!} g_n^2 (P_n^{(s)}(\mu))^2.$$

But for $\mu \in [-1, 1]$, $-n \leq s \leq n$ we have

$$|P_n^{(s)}(\mu)| \leqslant \frac{(n+|s|)!}{n!},^{1)}$$

which implies immediately that

$$[\sigma_{N_1 N_2}]^2 \leqslant \sum_{n=N_1}^{N_2} \frac{2n+1}{2} g_n^2 \frac{(n+|s|)!\,(n-|s|)!}{(n!)^2}$$

for all $\mu \in [-1, 1]$.

If $s \neq 0$, then

$$\frac{(n+|s|)!\,(n-|s|)!}{(n!)^2} = \prod_{j=1}^{|s|} \frac{1+j/n}{1-\frac{j-1}{n}} \leqslant \prod_{j=1}^{|s|} \frac{1+j|s|^{-1}}{1-\frac{j-1}{|s|}}$$

$$\leqslant \prod_{j=1}^{|s|} \frac{2}{1-\frac{|s|-1}{|s|}} = (2|s|)^{|s|}$$

and

$$[\sigma_{N_1 N_2}]^2 \leqslant (2|s|)^{|s|} \sum_{n=N_1}^{N_2} \frac{2n+1}{2} g_n^2, \quad \mu \in [-1,1].$$

All that is necessary to complete the proof of the lemma is to refer to Theorem 2.

Lemma 11. *Let s be an integer. There exists a function $h_s(\mu, \mu')$ which is quadratically summable on D and has the following properties:*

[i] *$h_s(\mu, \mu')$ is the sum of series (38),*

$$\lim_{N=\infty} \int_D \left[h_s(\mu, \mu') - \sum_{n=|s|}^{N} \frac{2n+1}{2} \frac{(n-|s|)!}{(n+|s|)!} g_n P_n^{(s)}(\mu) P_n^{(s)}(\mu')\right]^2 d\mu\, d\mu' = 0.$$

[ii] *Being a function of μ', $h_s(\mu, \mu')$ is an element of $L_2(-1, 1)$ for every fixed $\mu \in [-1, 1]$. Here*

$$\lim_{N=\infty} \int_{-1}^{1} \left[h_s(\mu, \mu') - \sum_{n=|s|}^{N} \frac{2n+1}{2} \frac{(n-|s|)!}{(n+|s|)!} g_n P_n^{(s)}(\mu) P_n^{(s)}(\mu')\right]^2 d\mu' = 0$$

uniformly with respect to all $\mu \in [-1, 1]$.

[iii] *Almost everywhere on D we have*

1) This estimate is readily derivable, for example, from formula (8.816) of [37].

$$h_s(\mu, \mu') = \int_0^{2\pi} g(\gamma_0(\mu, \mu', \varphi))\, e^{is\varphi} d\varphi.$$

Proof. According to Lemma 10, series (38) converges to some function $h'_s \in L_2(D)$ in the metric of $L_2(D)$; it converges to the function $h''_s(\mu, \mu')$ in the metric of $L_2(-1, 1)$ for every fixed μ. The function $h''_s(\mu, \mu')$ is quadratically summable over its second argument on $[-1, 1]$. Moreover, the limit relation

$$\lim_{N=\infty} \int_{-1}^{1} \left[h''_s(\mu, \mu') - \sum_{n=|s|}^{N} \frac{2n+1}{2} \frac{(n-|s|)!}{(n+|s|)!} g_n P_n^{(s)}(\mu) P_n^{(s)}(\mu') \right]^2 d\mu' = 0$$

is fulfilled uniformly with respect to all $\mu \in [-1, 1]$.

By virtue of the Fubini theorem, there exists an $\mathfrak{S} \subset [-1, 1]$ such that $\operatorname{mes} \mathfrak{S} = 2$; for $\mu \in \mathfrak{S}$ the function $h'_s(\mu, \mu')$ is quadratically summable over μ' on $[-1, 1]$. Let us choose an integer $n_0 \geq |s|$ and set

$$f(\mu) = \int_{-1}^{1} h'_s(\mu, \mu') P_{n_0}^{(s)}(\mu') d\mu' - g_{n_0} P_{n_0}^{(s)}(\mu)$$

for $\mu \in \mathfrak{S}$. It is clear that $f \in L_2(-1, 1)$. Let $k \geq |s|$. We have

$$\int_{-1}^{1} P_k^{(s)}(\mu) f(\mu) d\mu$$

$$= \int_D h'_s(\mu, \mu') P_k^{(s)}(\mu) P_{n_0}^{(s)}(\mu') d\mu' d\mu - g_{n_0} \int_{-1}^{1} P_{n_0}^{(s)}(\mu) P_k^{(s)}(\mu) d\mu$$

$$= \sum_{n=|s|}^{\infty} \frac{2n+1}{2} \frac{(n-|s|)!}{(n+|s|)!} g_n \left(\frac{(n+|s|)!}{(n-|s|)!} \cdot \frac{2}{2n+1} \delta_{nk} \right) \left(\frac{(n+|s|)!}{(n-|s|)!} \cdot \frac{2}{2n+1} \delta_{nn_0} \right)$$
$$- g_{n_0} \frac{(n_0+|s|)!}{(n_0-|s|)!} \cdot \frac{2}{2n_0+1} \delta_{kn_0} = 0.$$

The system of functions $\{P_k^{(s)} \mid k \geq |s|\}$ is complete in $L_2(-1, 1)$. Hence $f(\mu) = 0$ almost everywhere on $\mathfrak{S}$. Eliminating a set of measure zero from $\mathfrak{S}$ if necessary, we conclude that for every $\mu \in \mathfrak{S}$ we have

$$\int_{-1}^{1} h'_s(\mu, \mu') P_k^{(s)}(\mu') d\mu' = g_k P_k^{(s)}(\mu), \quad k \geqslant |s|.$$

Let $\mu \in [-1, 1]$. By the definition of $h''_s(\mu, \mu')$, for $k \geq |s|$ we have

$$\int_{-1}^{1} h''_s(\mu, \mu') P_k^{(s)}(\mu') d\mu'$$

$$= \sum_{n=|s|}^{\infty} \frac{2n+1}{2} \frac{(n-|s|)!}{(n+|s|)!} g_n P_n^{(s)}(\mu) \left(\frac{(n+|s|)!}{(n-|s|)!} \cdot \frac{2}{2n+1} \delta_{nk} \right) = g_k P_k^{(s)}(\mu).$$

It is now clear that the equation

$$h'_s(\mu, \mu') = h''_s(\mu, \mu')$$

is fulfilled for almost all $\mu' \in [-1, 1]$ for every $\mu \in \mathscr{E}$.

Let us set

$$h_s(\mu, \mu') = \begin{cases} h_s'(\mu, \mu') & \text{for } \mu \in \mathscr{E},\ \mu' \in [-1,1]; \\ h_s''(\mu, \mu') & \text{for } \mu \in [-1,1] \setminus \mathscr{E},\ \mu' \in [-1,1]. \end{cases}$$

This definition implies that $h_s(\mu, \mu') = h_s'(\mu, \mu')$ almost everywhere on D. Hence $h_s \in L_2(D)$, and h_s is the sum (in the metric of $L_2(D)$) of series (38). Let us choose a $\mu \in [-1, 1]$. If $\mu \in \mathscr{E}$, then $h_s(\mu, \mu') = h_s'(\mu, \mu') = h_s''(\mu, \mu')$ for almost all $\mu' \in [-1, 1]$. If $\mu \,\overline{\in}\, \mathscr{E}$, then, by definition, $h_s(\mu, \mu') = h_s''(\mu, \mu')$ for all $\mu' \in [-1, 1]$. Thus $h_s(\mu, \mu')$ is quadratically summable over μ' on $[-1, 1]$ for every $\mu \in [-1, 1]$. Here

$$\int_{-1}^{1} \Big[h_s(\mu, \mu') - \sum_{n=|s|}^{N} \frac{2n+1}{2} \frac{(n-|s|)!}{(n+|s|)!} g_n P_n^{(s)}(\mu) P_n^{(s)}(\mu')\Big]^2 d\mu'$$

$$= \int_{-1}^{1} \Big[h_s''(\mu, \mu') - \sum_{n=|s|}^{N} \frac{2n+1}{2} \frac{(n-|s|)!}{(n+|s|)!} g_n P_n^{(s)}(\mu) P_n^{(s)}(\mu')\Big]^2 d\mu' \to 0$$

as $N \to \infty$ uniformly with respect to $\mu \in [-1, 1]$. It remains for us to prove[iii].

We set

$$u_s(\mu, \mu') = \int_0^{2\pi} g(\gamma_0(\mu, \mu', \varphi))\, e^{is\varphi} d\varphi$$

for $(\mu, \mu') \in D$. It follows from Lemma 9 and from the Fubini theorem that $u_s \in L_2(D)$. For $n \geq |s|$, $n' \geq |s|$ we have

$$\int_D u_s(\mu, \mu') P_n^{(s)}(\mu) P_{n'}^{(s)}(\mu')\, d\mu d\mu'$$

$$= \int_{-1}^{1} P_n^{(s)}(\mu)\, d\mu \int_\Delta g(\gamma_0(\mu, \mu', \varphi'))\, e^{is\varphi'} P_{n'}^{(s)}(\mu')\, d\mu' d\varphi'$$

$$= \Big[\frac{4\pi}{2n'+1} \frac{(n'+|s|)!}{(n'-|s|)!}\Big]^{1/2} \int_{-1}^{1} P_n^{(s)}(\mu)\, d\mu \int_\Delta g(\gamma_0(\mu, \mu', \varphi'))\, Y_{n'}^{(s)}(\mu', \varphi')\, d\mu' d\varphi'.$$

Let σ be an orthonormal basis of R_3, and let $\omega = \psi_\sigma(\mu, 0)$. Then (compare Lemmas 2 and 1 and Theorem 2)

$$\int_\Delta g(\gamma_0(\mu, \mu', \varphi'))\, Y_{n'}^{(s)}(\mu', \varphi')\, d\mu' d\varphi' = \int_\Omega g(\omega\omega')\, Y_{\sigma n'}^{(s)}(\omega')\, d\omega'$$

$$= \hat{g} Y_{\sigma n'}^{(s)}(\omega) = g_{n'} Y_{\sigma n'}^{(s)}(\omega) = g_{n'} Y_{n'}^{(s)}(\mu, 0) = g_{n'} \Big[\frac{2n'+1}{4\pi} \frac{(n'-|s|)!}{(n'+|s|)!}\Big]^{1/2} P_{n'}^{(s)}(\mu).$$

Hence

$$\int_D u_s(\mu, \mu') P_n^{(s)}(\mu) P_{n'}^{(s)}(\mu')\, d\mu d\mu' = g_{n'} \int_{-1}^{1} P_n^{(s)}(\mu) P_{n'}^{(s)}(\mu)\, d\mu$$

$$= \frac{2}{2n+1} \frac{(n+|s|)!}{(n-|s|)!} g_n \delta_{nn'}, \quad n \geqslant |s|, \quad n' \geqslant |s|.$$

At the same time, by [i],

$$\int_D h_s(\mu,\mu')P_n^{(s)}(\mu)P_{n'}^{(s)}(\mu')\,d\mu d\mu'$$

$$=\sum_{k=|s|}^{\infty}\frac{2k+1}{2}\,\frac{(k-|s|)!}{(k+|s|)!}\,g_k\int_{-1}^{1}P_k^{(s)}(\mu)P_n^{(s)}(\mu)\,d\mu\int_{-1}^{1}P_k^{(s)}(\mu')P_{n'}^{(s)}(\mu')\,d\mu'$$

$$=\frac{2}{2n+1}\,\frac{(n+|s|)!}{(n-|s|)!}\,g_n\delta_{nn'},\quad n\geqslant|s|,\quad n'\geqslant|s|.$$

Since the system $\{P_n^{(s)}(\mu)\,P_{n'}^{(s)}(\mu')\;|n\geq|s|,\,n'\geq|s|\}$ is complete in $L_2(D)$, we find that $h_s(\mu,\mu')=u_s(\mu,\mu')$ almost everywhere on D. Lemma 11 is proved.

To each function $h_s(\mu,\mu')$, $s=0,\pm1,\pm2,\cdots$, we assign an operator $\hat{h}_s\colon L_2(-1,1)\to L_2(-1,1)$ according to the rule

$$\hat{h}_s f(\mu)=\int_{-1}^{1}h_s(\mu,\mu')f(\mu')\,d\mu',\qquad f\in L_2(-1,1).$$

It is clear that $\hat{h}_s$ is a linear completely continuous selfadjoint operator.

Lemma 12. *Let $a(\mu,\phi)$ be a function quadratically summable on Δ; let σ be an orthonormal basis of R_3. We set (compare §2)*

$$a_\sigma(\omega)=\begin{cases}a(\mu,\varphi), & \text{if}\quad \omega\in\Omega\setminus M_\sigma\ u\ (\mu,\varphi)\in\psi_\sigma^{-1}(\omega),\\ 0, & \text{if}\quad \omega\in M_\sigma.\end{cases}$$

This relation defines the function a_σ uniquely, and $a_\sigma\in L_2(\Omega)$. If $(\mu,\phi)\in\Delta$, $\omega=\psi_\sigma(\mu,\phi)$, then $\hat{g}a_\sigma(\omega)=\int_\Delta g(\gamma(\mu,\phi,\mu',\phi'))\,a(\mu',\phi')\,d\mu'd\phi'$.

Specifically, if $a(\mu,\phi)=e^{is\phi}q(\mu)$, where s is an integer and $q\in L_2(-1,1)$, then $\hat{g}a_\sigma(\omega)=e^{is\phi}\hat{h}_s q(\mu)$, $(\mu,\phi)\in\Delta$ and $\omega=\psi_\sigma(\mu,\phi)$.

Proof. The unique definition of $a_\sigma(\omega)$ follows from the fact that $\psi_\sigma^{-1}(\omega)$, $\omega\in\Omega\setminus M_\sigma$, contains only one point (μ,ϕ) (compare §2). We note that

$$a_\sigma(\psi_\sigma(\mu,\varphi))=\begin{cases}a(\mu,\varphi) & \text{for}\ (\mu,\varphi)\in\Delta_0;\\ 0 & \text{for}\ (\mu,\varphi)\in\Delta\setminus\Delta_0.\end{cases}$$

In accordance with Lemma 1 this implies that $a_\sigma\in L_2(\Omega)$. Let $(\mu,\phi)\in\Delta$, $\omega=\psi_\sigma(\mu,\phi)$. We have

$$\hat{g}a_\sigma(\omega)=\int_\Omega g(\omega\omega')\,a_\sigma(\omega')\,d\omega'=\int_\Delta g(\omega\psi_\sigma(\mu',\varphi'))\,a_\sigma(\psi_\sigma(\mu',\varphi'))\,d\mu'd\varphi'$$

$$=\int_\Delta g(\gamma(\mu,\varphi,\mu',\varphi'))\,a(\mu',\varphi')\,d\mu'd\varphi'.$$

If $a(\mu,\phi)=e^{is\phi}q(\mu)$, if s is an integer, and if $q\in L_2(-1,1)$, then, by Lemma 11,

$$\hat{g}a_\sigma(\omega) = \int_{-1}^{1} q(\mu')d\mu' \int_0^{2\pi} g(\gamma_0(\mu, \mu', \varphi' - \varphi))\, e^{is\varphi'}d\varphi'$$

$$= e^{is\varphi} \int_{-1}^{1} q(\mu')d\mu' \int_0^{2\pi} g(\gamma_0(\mu, \mu', \varphi'))\, e^{is\varphi'}d\varphi' = e^{is\varphi}\hat{h}_s q(\mu).$$

Theorem 13. *Let* $\mathrm{k} \in Z_0$, $b \in L_2(-1, 1)$, *let* λ *be a number, and let*

$$\lambda(1 + \mathrm{k}\mu)\, b(\mu) = \hat{h}_s b(\mu), \quad \mu \in [-1, 1]$$

for some number s. *We set* $a(\mu, \phi) = e^{is\phi}b(\mu)$, $(\mu, \phi) \in \Delta$, *and let* σ *be an orthonormal basis of the space* R_3 *with the third unit vector* $\mathbf{n}$. *We assign to the function* $a(\mu, \phi)$ *a function on the sphere* $a_\sigma(\omega)$ *as in Lemma* 12. *Then*

$$\hat{U}(\mathrm{k})\, a_\sigma(\omega) = \lambda a_\sigma(\omega)$$

almost everywhere on Ω.

Proof. Let $\omega \in \Omega$. We choose a $(\mu, \phi) \in \psi_\sigma^{-1}(\omega)$. Then $(\mu, \phi) \in \Delta$, $\psi_\sigma(\mu, \phi) = \omega$. By Lemma 12,

$$\hat{g}a_\sigma(\omega) = e^{is\varphi}\hat{h}_s b(\mu) = \lambda(1 + \mathrm{k}\mu)\, b(\mu)\, e^{is\varphi} = \lambda(1 + \mathrm{k}\omega\mathbf{n})\, a(\mu, \varphi).$$

If $\omega \,\overline{\in}\, M_\sigma$, then $(\mu, \phi) \in \Delta_0$ and $a_\sigma(\omega) = a_\sigma(\psi_\sigma(\mu, \phi)) = a(\mu, \phi)$. This means that

$$\hat{g}a_\sigma(\omega) = \lambda(1 + \mathrm{k}\omega\mathbf{n})\, a_\sigma(\omega)$$

for $\omega \in \Omega \backslash M_\sigma$, and the theorem is proved.

Theorem 13 enables us to construct the eigenfunctions of the operator $\hat{U}(\mathrm{k})$ with separated variables. It turns out that the operator $\hat{U}(\mathrm{k})$ admits of no other eigenfunctions.

Theorem 14. *Let* $\mathrm{k} \in Z_0$ *and* $\lambda \in \sigma(\hat{U}(\mathrm{k}))$, $\lambda \neq 0$. *If* $f \in L_2(\Omega)$, $f \neq \vartheta$, $\hat{U}(\mathrm{k})f = \lambda f$, *then the functions* $f_s = \hat{P}^{(s)}_{\mathbf{n}} f$, $s = 0, \pm 1, \pm 2, \cdots$, *include only a finite number of nonzero functions* f_{s_ν}, $\nu = 1, \cdots, \nu_0$. *Here*

$$\hat{U}(\mathrm{k})\, f_{s_\nu} = \lambda f_{s_\nu}, \; \nu = 1, 2, \ldots, \nu_0$$

and

$$f = \sum_{\nu=1}^{\nu_0} f_{s_\nu}.$$

Let σ *be an orthonormal basis of* R_3 *with the third unit vector* $\mathbf{n}$. *There exist functions* $a_\nu \in C(-1, 1)$ *such that*

$$f_{s_\nu}(\omega) = a_\nu(\mu)\, e^{is_\nu\varphi}, \quad (\mu, \varphi) \in \psi_\sigma^{-1}(\omega)$$

for all $\omega \in \Omega$. *The function* $a_\nu(\mu)$ *is uniquely defined for each* ν, $1 \leq \nu \leq \nu_0$. *It also satisfies the equation*

$$\lambda(1 + k\mu)\, a_\nu(\mu) = \hat{h}_{s_\nu} a_\nu(\mu), \quad \mu \in [-1, 1].$$

Proof. Let $f \in L_2(\Omega)$, $\hat{U}(k) f = \lambda f$, $f \neq \vartheta$. By Theorem 1, $f \in C(\Omega)$. The function

$$\lambda(1 + k\mu)$$

of the variable μ is continuous on $[-1, 1]$. Making use of Lemma 3 and Theorem 2, we obtain

$$\lambda(1 + k\omega\mathbf{n}) f_s(\omega) = \lambda(1 + k\omega\mathbf{n}) \hat{P}^{(s)}_{\mathbf{n}} f(\omega)$$
$$= \hat{P}^{(s)}_{\mathbf{n}} [\lambda(1 + k\omega\mathbf{n}) f(\omega)] = \hat{P}^{(s)}_{\mathbf{n}} \hat{g} f(\omega) = \hat{g}\hat{P}^{(s)}_{\mathbf{n}} f(\omega) = \hat{g} f_s(\omega)$$

and

$$\hat{U}(k) f_s(\omega) = \lambda f_s(\omega), \quad s = 0, \pm 1, \pm 2, \ldots.$$

Thus every nonzero function f_s is an eigenfunction of $\hat{U}(k)$ which corresponds to the same eigenvalue $\lambda \neq 0$. The eigensubspace of the operator $\hat{U}(k)$ belonging to λ is finite-dimensional, and the system of all nonzero f_s is linearly independent (this follows from the orthogonality of the projective operators $\hat{P}^{(s)}_{\mathbf{n}}$). This implies that only a finite number of functions f_s can be nonzero. Let us number these nonzero functions to form a finite sequence, $\{f_{s_\nu} \mid \nu = 1, \cdots, \nu_0\}$. It is clear that $f = \sum_{\nu=1}^{\nu_0} f_{s_\nu}$.

Let us isolate an orthonormal basis σ of the space R_3 with the third unit vector $\mathbf{n}$. Let

$$a_\nu(\mu) = \frac{1}{2\pi} \int_0^\infty e^{-is_\nu\varphi'} f(\psi_\sigma(\mu, \varphi'))\, d\varphi', \quad \nu = 1, 2, \ldots, \nu,$$

for $\mu \in [-1, 1]$. Clearly, $a_\nu \in C(-1, 1)$. According to Lemma 3,

$$f_{s_\nu}(\omega) = \hat{P}^{(s_\nu)}_{\mathbf{n}} f(\omega) = e^{is_\nu\varphi} a_\nu(\mu), \quad \nu = 1, 2, \ldots, \nu_0.$$

We choose a ν and set $a(\mu, \phi) = e^{is_\nu\phi} a_\nu(\mu)$, $(\mu, \phi) \in \Delta$. We assign to the function $a(\mu, \phi)$ a function $a_\sigma(\omega)$, $\omega \in \Omega$ (Lemma 12). Let $\omega \in \Omega \setminus M_\sigma$ and $(\mu, \phi) \in \psi_\sigma^{-1}(\omega)$. By the definition of a_σ,

$$a_\sigma(\omega) = a(\mu, \varphi) = f_{s_\nu}(\omega).$$

Thus $a_\sigma(\omega) = f_{s_\nu}(\omega)$ almost everywhere on Ω. This means that $\hat{g} f_{s_\nu}(\omega) = \hat{g} a_\sigma(\omega)$ everywhere on Ω.

For $(\mu, \phi) \in \Delta$ we set $\omega = \psi_\sigma(\mu, \phi)$. Then $\lambda(1 + k\mu) e^{is_\nu\phi} a_\nu(\mu) =$

$\lambda(1 + k\omega\mathbf{n}) f_{s\nu}(\omega) = \hat{g} f_{s\nu}(\omega) = \hat{g} a_\sigma(\omega) = e^{is\nu\phi} \hat{h}_{s\nu} a_\nu(\mu)$ (compare Lemma 12). The theorem is proved.

We shall use Theorem 14 in order to establish the fact that the eigenfunction Φ_k of the operator $\hat{U}(k)$, $k \in (-1, 1)$ (Theorem 6) is independent of the azimuth angle. First, we note the following lemma.

Lemma 13. *Let* $F \in L_2(\Omega)$, $\hat{P}^{(0)}_{\mathbf{n}} F = F$. *There then exists an* $f \in L_2(-1, 1)$ *such that* $F(\omega) = f(\omega\mathbf{n})$ *for almost all* $\omega \in \Omega$. *If, in addition,* $F \in C(\Omega)$, *then* $f \in C(-1, 1)$.

Proof. Let us choose an orthonormal basis σ of the space R_3 with the third unit vector $\mathbf{n}$. If $F \in C(\Omega)$, then for every $\omega \in \Omega$ (compare Lemma 3) we have

$$F(\omega) = \hat{P}^{(0)}_{\mathbf{n}} F(\omega) = \frac{1}{2\pi} \int_0^{2\pi} F(\psi_\sigma(\mu, \varphi'))\, d\varphi' \equiv f(\mu),\ (\mu, \varphi) \in \psi_\sigma^{-1}(\omega).$$

Clearly, $f \in C(-1, 1)$, and since $\mu = \omega\mathbf{n}$ for $(\mu, \phi) \in \psi_\sigma^{-1}(\omega)$, it follows that $F(\omega) = f(\omega\mathbf{n})$ for all $\omega \in \Omega$.

Now let $F \in L_2(\Omega)$. We choose a sequence $\{F_k | k = 1, 2, \cdots\} \subset C(\Omega)$ such that $F_k \to F$ in the metric of $L_2(\Omega)$ as $k \to \infty$. Each function F_k is associated with a function $f_k \in C(-1, 1)$ such that $F_k(\omega) = f_k(\omega\mathbf{n})$, $\omega \in \Omega$.

$$\| F_k - F_{k'} \|^2 = \int_\Omega | f_k(\omega\mathbf{n}) - f_{k'}(\omega\mathbf{n}) |^2 \, d\omega = 2\pi\, [f_k - f_{k'}]^2.$$

This implies that $\{f_k | k = 1, 2, \cdots\}$ is fundamental in $L_2(-1, 1)$, which means that the limit $f = \lim_{k=\infty} f_k \in L_2(-1, 1)$ exists. We have

$$\| F(\omega) - f(\omega\mathbf{n}) \| \leqslant \| F - F_k \| + \| f_k(\omega\mathbf{n}) - f(\omega\mathbf{n}) \|$$
$$= \| F - F_k \| + \sqrt{2\pi}\, [f_k - f], \quad k = 1, 2, \ldots.$$

Hence $\|F(\omega) - f(\omega\mathbf{n})\| = 0$, and $F(\omega) = f(\omega\mathbf{n})$ almost everywhere on Ω.

Theorem 15. *Let us use the notation of Theorem 7. Then* $\hat{P}^{(s)}_{\mathbf{n}} e_0(k) = \delta_{s0} e_0(k)$, $k \in G_0$, $s = 0, \pm 1, \pm 2, \cdots$. *There exists a function* ϕ_k *with values in* $C(-1, 1)$ *which is defined for* $k \in G_0$ *and has the following properties:*

[i] ϕ_k *is a holomorphic function of* $k \in G_0$.

[ii] *For* $k \in G_0$ *and* $\omega \in \Omega$ *we have*

$$e_0(k, \omega) = \varphi_k(\omega\mathbf{n}).$$

Specifically,

$$\Phi_k(\omega) = \varphi_k(\omega\mathbf{n}), \quad \omega \in \Omega, \quad k \in (-1, 1).$$

[iii] *For* $k \in (-1, 1)$ *the function* ϕ_k *satisfies the equation*

$$M(k)(1 + k\mu)\, \varphi_k(\mu) = \hat{h}_0 \varphi_k(\mu), \quad \mu \in (-1, 1)$$

and the normalization condition

$$\int_{-1}^{1} \varphi_k(\mu)\, d\mu = 2.$$

The equations

$$M(k)(1 + k\mu)\, x(\mu) = \hat{h}_s x(\mu)$$

admit of only the trivial solution x, $[x] = 0$ *in* $L_2(-1, 1)$ *for* $s \neq 0$, $k \in (-1, 1)$.

Proof. Let $k \in (-1, 1)$. Then $\hat{U}(k)\Phi_k = M(k)\Phi_k$, and the eigensubspace of the operator $\hat{U}(k)$ belonging to the eigenvalue $M(k)$ is one-dimensional (Theorem 6). This subspace contains all of the vectors $\hat{P}^{(s)}_{n}\Phi_k$, $s = 0, \pm 1, \pm 2, \cdots$, (Theorem 14). This means that only one of these vectors can be nonzero. We choose an orthonormal basis σ of the space R_3 with the third unit vector n. For $\omega \in \Omega$, $(\mu, \phi) \in \psi_\sigma^{-1}(\omega)$ we have

$$\hat{P}^{(0)}_{n}\Phi_k(\omega) = \frac{1}{2\pi}\int_0^{2\pi} \Phi_k(\psi_\sigma(\mu, \varphi'))\, d\varphi' > 0,$$

which means that $\hat{P}^{(0)}_{n}\Phi_k \neq \vartheta$, $k \in (-1, 1)$. According to the above, all of the other projections of Φ_k are equal to zero, i.e. $\hat{P}^{(s)}_{n}\Phi_k = \vartheta$, $s = \pm 1, \pm 2, \cdots$, $k \in (-1, 1)$. Thus $\Phi_k = \hat{P}^{(0)}_{n}\Phi_k$, $k \in (-1, 1)$, which means that

$$\hat{P}^{(s)}_{n}\Phi_k = \delta_{s0}\Phi_k, \qquad s = 0, \pm 1, \pm 2, \ldots, k \in (-1, 1).$$

Clearly, $\hat{P}^{(s)}_{n} e_0(k)$ is a function of k holomorphic to G_0 for any integer s. For $k \in (-1, 1)$ it coincides with $\delta_{s0} e_0(k)$:

$$\hat{P}^{(s)}_{n} e_0(k) = \hat{P}^{(s)}_{n}\Phi_k = \delta_{s0}\Phi_k = \delta_{s0} e_0(k), \quad k \in (-1, 1).$$

Since $\delta_{s0} e_0(k)$ is analytic in G_0, the equation

$$\hat{P}^{(s)}_{n} e_0(k) = \delta_{s0} e_0(k), \quad s = 0, \pm 1, \pm 2, \ldots$$

remains valid for all $k \in G_0$. Specifically (compare Theorem 8),

$$\hat{P}^{(0)}_{n} e_0^{[q]}(k) = e_0^{[q]}(k) \in C(\Omega), \quad k \in G_0, \quad q = 0, 1, 2, \ldots.$$

By Lemma 13, for each $k \in G_0$ we have uniquely defined functions $\phi_k^{[q]} \in C(-1, 1)$ such that

$$e_0^{[q]}(k, \omega) = \varphi_k^{[q]}(\omega n), \quad \omega \in \Omega, \quad k \in G_0, \quad q = 0, 1, 2, \ldots.$$

Let us set $\phi_k = \phi_k^{[0]}$, $k \in G_0$. Then

$$e_0(k, \omega) = \varphi_k(\omega n), \quad \omega \in \Omega, \quad k \in G_0;$$
$$\Phi_k(\omega) = \varphi_k(\omega n), \quad \omega \in \Omega, \quad k \in (-1, 1).$$

Let $k_1 \in G_0$. According to Theorem 8, there exists an $r > 0$ such that for $|k - k_1| < r$ we have

$$e_0(k, \omega) = \sum_{q=0}^{\infty} \frac{1}{q!}(k - k_1)^q e_0^{[q]}(k_1, \omega),$$

where the series on the right converges uniformly with respect to all $(k, \omega) \in S(k_1, r) \times \Omega$. The above equation can now be rewritten as

$$\varphi_k(\omega\mathbf{n}) = \sum_{q=0}^{\infty} \frac{1}{q!}(k - k_1)^q \varphi_{k_1}^{[q]}(\omega\mathbf{n}).$$

This means that

$$\varphi_k(\mu) = \sum_{q=0}^{\infty} \frac{1}{q!}(k - k_1)^q \varphi_{k_1}^{[q]}(\mu),$$

uniformly with respect to all $(k, \mu) \in S(k_1, r) \times [-1, 1]$, or, which is the same thing, that

$$\varphi_k = \sum_{q=0}^{\infty} \frac{1}{q!}(k - k_1)^q \varphi_{k_1}^{[q]}.$$

The series on the right converges uniformly in the metric of $C(-1, 1)$ with respect to $k \in S(k_1, r)$. Thus the function ϕ_k with values in $C(-1, 1)$ is regular at the point $k = k_1$. Since $k_1 \in G_0$ is arbitrary, ϕ_k is a holomorphic function of k in the domain G_0.

It remains for us to prove [iii]. Let $k \in (-1, 1)$. We note that

$$\hat{P}_{\mathbf{n}}^{(0)} \Phi_k(\omega) = \Phi_k(\omega) = \varphi_k(\mu) e^{i0\varphi}, \quad (\mu, \varphi) \in \psi_\sigma^{-1}(\omega)$$

for $\omega \in \Omega$. From Theorem 14 we now infer that $\phi_k(\mu)$ satisfies the equation

$$M(k)(1 + k\mu)\varphi_k(\mu) = \hat{h}_0 \varphi_k(\mu), \quad \mu \in [-1, 1].$$

The function Φ_k is normalized by the condition $(\Phi_k, 1) = 4\pi$. Hence

$$\int_{-1}^{1} \varphi_k(\mu)\, d\mu = \frac{1}{2\pi} \int_{\Omega} \varphi_k(\omega\mathbf{n})\, d\omega = \frac{1}{2\pi}(\Phi_k, 1) = 2.$$

Finally, let us suppose that $k \in (-1, 1)$, $x \in L_2(\Omega)$, and

$$M(k)(1 + k\mu)x(\mu) = \hat{h}_s x(\mu), \quad s \neq 0.$$

Let $a(\mu, \phi) = x(\mu)e^{is\phi}$, $(\mu, \phi) \in \Delta$. Let us associate the function $a(\mu, \phi)$ with a function on the sphere $a_\sigma(\omega)$ (compare Lemma 12). According to Theorem 13,

$$\hat{U}(\mathrm{k})\, a_\sigma = M(\mathrm{k})\, a_\sigma.$$

Since $M(\mathrm{k})$ is a simple eigenvalue of $\hat{U}(\mathrm{k})$, there exists a number κ such that $a_\sigma = \kappa\Phi_{\mathrm{k}}$,

$$4\pi\varkappa = \varkappa(\Phi_{\mathrm{k}}, 1) = (a_\sigma, 1).$$

But

$$(a_\sigma, 1) = \int_\Delta a_\sigma(\psi_\sigma(\mu, \varphi))\, d\mu d\varphi = \int_{-1}^{1} x(\mu)\, d\mu \int_0^{2\pi} e^{is\varphi} d\varphi = 0.$$

Hence $\kappa = 0$, which means that $a_\sigma = \vartheta$. We have

$$0 = \|a_\sigma\|^2 = \int_\Omega |a_\sigma(\omega)|^2\, d\omega = \int_\Delta |a_\sigma(\psi_\sigma(\mu, \varphi))|^2\, d\mu d\varphi$$
$$= \int_\Delta |x(\mu)|^2\, d\mu d\varphi = 2\pi\, [x]^2,$$

and the theorem is proved.

§6. The inhomogeneous characteristic equation

Here we shall investigate the equation

$$(1 + \mathrm{k}\omega\mathrm{n})\, \psi(\mathrm{k}, \omega) = \hat{g}\psi(\mathrm{k}, \omega) + \alpha(\mathrm{k}, \omega), \tag{39}$$

whose absolute term $\alpha(\mathrm{k})$ is a function defined on the sphere Ω with the values $\alpha(\mathrm{k}, \omega)$, $\omega \in \Omega$; $\alpha(\mathrm{k}) \in L_2(\Omega)$ for every permissible k. We shall be interested primarily in the character of the dependence of the solution $\psi(\mathrm{k})$ of equation (39) on the complex parameter k. Under natural assumptions concerning $\alpha(\mathrm{k})$ the solution $\psi(\mathrm{k})$ turns out to be a holomorphic function of k with singularities at the points $\mathrm{k} \in \mathfrak{N}$. The form of these singularities is determined by the solutions of the homogeneous characteristic equation for the corresponding values of the parameter $\mathrm{k} \in \mathfrak{N}$. For every $\mathrm{k} \in \mathfrak{N}$ the solutions of the homogeneous equation form a finite-dimensional subspace (Lemma 14) which admits of a basis of special form (Lemma 15). The principal terms of the expansions of $\psi(\mathrm{k})$ in Laurent series in the neighborhoods of the singular points can be expressed particularly simply in terms of the elements of these bases. The possibility of such expansions and the very fact that $\psi(\mathrm{k})$ is analytic follow from the mode of construction of the coefficients of these expansions (Lemmas 16, 17, and 18). The basic result of the present section is contained in Theorem 16.

As before, we assume that the indicatrix g and the vector $\mathrm{n} \in \Omega$ are fixed.

We define the operator $\hat{W}(k)$ for $k \in Z_0$ by the expression

$$\hat{W}(k) f(\omega) = (1 + k\omega n) f(\omega) - \hat{g} f(\omega), \quad f \in L_2(\Omega).$$

It is clear that $\hat{W}(k)$ is a linear bounded operator acting from $L_2(\Omega)$ into $L_2(\Omega)$.

Lemma 14. *For $k \in Z_0 \setminus \mathfrak{N}$ the operator $\hat{W}(k)$ effects a one-to-one mapping of $L_2(\Omega)$ onto itself. Choose a $k \in \mathfrak{N}$. The set*

$$H_k = \{f \mid f \in L_2(\Omega), \quad \hat{W}(k) f = \vartheta\}$$

is a subspace of $L_2(\Omega)$ of dimensionality p_k, $1 \le p_k < \infty$. We denote by $\mathfrak{L}_k$ the orthogonal complement of H_k in $L_2(\Omega)$. Then $\hat{W}(k)\,(L_2(\Omega)) \subset \mathfrak{L}_k$. Let $\hat{N}_k$ be an operator induced in $\mathfrak{L}_k$ by the operator $\hat{W}(k)$,

$$\hat{N}_k f = \hat{W}(k) f, \quad f \in \mathcal{L}_k.$$

$\hat{N}_k$ is a linear bounded selfadjoint operator which effects a one-to-one mapping of $\mathfrak{L}_k$ onto $\mathfrak{L}_k$.

Proof. For $k \in Z_0$ and $f \in L_2(\Omega)$ we have

$$\hat{W}(k) f = -(1 + k\omega n)(\hat{U}(k) - \hat{I}) f.$$

If $k \bar{\in} \mathfrak{N}$, then the number 1 is a regular value of the operator $\hat{U}(k)$. This implies that $\hat{W}(k)$ effects a one-to-one mapping of $L_2(\Omega)$ onto itself, $k \bar{\in} \mathfrak{N}$.

Let $k \in \mathfrak{N}$. Then 1 is an eigenvalue of the completely continuous operator $\hat{U}(k)$, and H_k is the corresponding eigensubspace. Hence H_k is finite-dimensional and contains at least one nonzero element, which means that $1 \le p_k < \infty$. It is clear that $\hat{W}(k)$ is a selfadjoint operator for $k \in \mathfrak{N}$, so that $\hat{W}(k)\,(L_2(\Omega)) \subset \mathfrak{L}_k$. We define $\hat{N}_k$ as we did in the lemma; we have $\hat{N}_k(\mathfrak{L}_k) \subset \mathfrak{L}_k$. The selfadjointness of $\hat{W}(k)$ implies that of $\hat{N}_k$. Let $\alpha \in \mathfrak{L}_k$. If $f \in L_2(\Omega)$ and $(\hat{U}(k) - \hat{I}) f = \vartheta$, then $f \in H_k$, which means that $(\alpha, f) = 0$. By Lemma 5, there exists a $\psi \in L_2(\Omega)$ such that $(\hat{U}(k) - \hat{I})\psi(\omega) = -\alpha(\omega)/(1 + k\omega n)$, or, which is the same thing, $\hat{W}(k)\psi = \alpha$. We denote by β the orthogonal projection of ψ onto $\mathfrak{L}_k$. Then $\beta \in \mathfrak{L}_k$, $\beta' = \psi - \beta \in H_k$, and $\hat{N}_k \beta = \hat{W}(k)\beta = \hat{W}(k)\psi - \hat{W}(k)\beta' = \alpha - \vartheta = \alpha$. Since α is an arbitrary element of $\mathfrak{L}_k$, we conclude that

$$\hat{N}_k(\mathcal{L}_k) = \mathcal{L}_k.$$

Finally, let $\alpha \in \mathfrak{L}_k$, $\hat{N}_k \alpha = \vartheta$. This means that $\hat{W}(k)\alpha = \vartheta$, so that

$\alpha \in H_k \cap \mathfrak{L}_k$. This means that $\alpha = \vartheta$, and $\hat{N}_k$ turns out to be a one-to-one mapping. The lemma is proved.

By virtue of this lemma and the Banach theorem on inverse operators, the operators $\hat{W}(k)$: $L_2(\Omega) \to L_2(\Omega)$ and $\hat{N}_k$: $\mathfrak{L}_k \to \mathfrak{L}_k$ have the linear bounded inverses $\hat{W}^{-1}(k)$ and $\hat{N}_k^{-1}$ for $k \in Z_0 \setminus \mathfrak{N}$ and $k \in \mathfrak{N}$, respectively.

We set

$$r(k) = \| \hat{W}^{-1}(k) \|, \; k \in Z_0 \setminus \mathfrak{N};$$
$$T(k) = \| \hat{N}_k^{-1} \|, \; k \in \mathfrak{N}.$$

Then $0 < r(k) < \infty$, $k \in Z_0 \setminus \mathfrak{N}$, and $0 < T(k) < \infty$, $k \in \mathfrak{N}$. The selfadjointness of $\hat{N}_k$ implies that $\hat{N}_k^{-1}$ is also a selfadjoint operator, $k \in \mathfrak{N}$.

The subspaces H_k corresponding to various values of the parameter $k \in \mathfrak{N}$ are orthogonal to each other with the weight (ωn). In fact, let $k_{1,2} \in \mathfrak{N}$, $k_1 \neq k_2$, $f_1 \in H_{k_1}$, $f_2 \in H_{k_2}$. Then

$$(1 + k_1 \omega n) f_1 = \hat{g} f_1,$$
$$(1 + k_2 \omega n) f_2 = \hat{g} f_2.$$

Let us take the scalar product of the right side of the first of these relations and f_2 and of the left side of the second relation and f_1,

$$(f_1, f_2) + k_1((\omega n) f_1, f_2) = (\hat{g} f_1, f_2),$$
$$(f_1, f_2) + k_2((\omega n) f_1, f_2) = (f_1, \hat{g} f_2).$$

The selfadjointness of f implies the equation $((\omega n) f_1, f_2) = 0$.

Among the elements of the set $\mathfrak{N}$ are the numbers $\pm\lambda_0$ (Theorem 10). The associated subspaces $H_{\pm\lambda_0}$ are one-dimensional (so that $p_{\pm\lambda_0} = 1$), and $\Phi_{\lambda_0} \in H_{\lambda_0}$, $\Phi_{-\lambda_0} \in H_{-\lambda_0}$. This follows from the fact that $H_{\pm\lambda_0}$ is the eigensubspace of the operator $\hat{U}(\pm\lambda_0)$ corresponding to the eigenvalue $M(\pm\lambda_0) = 1$, and from Theorem 6.

Lemma 15. $p_{-k} = p_k$ *for all* $k \in \mathfrak{N}$. *There exists a system of functions*

$$\{\psi_{kp} \mid k \in \mathfrak{N}, \; k \neq 0, \; p = 1, 2, \ldots, p_k\}$$

such that

[i] $\psi_{kp}(\omega)$ *is real*, $\omega \in \Omega$,

$$\psi_{kp} \in C(\Omega), \; p = 1, 2, \ldots, p_k, \; k \in \mathfrak{N}, \; k \neq 0.$$

[ii] $((\omega n)\psi_{kp}, \psi_{kp'}) = -\delta_{pp'} \operatorname{sgn} k$ *for* $1 \leq p \leq p' \leq p_k$, $k \in \mathfrak{N}$, $k \neq 0$.

[iii] $\psi_{(-k)p}(\omega) = \psi_{kp}(-\omega)$, $\omega \in \Omega$, $p = 1, 2, \cdots, p_k$, $k \in \mathfrak{N}$, $k \neq 0$.

[iv] $\{\psi_{kp} \mid p = 1, 2, \cdots, p_k\}$ *is a basis of* H_k, $k \in \mathfrak{N}$, $k \neq 0$.

Proof. Let us choose a $k \in \mathfrak{N}$ and let $\{t_p | p = 1, 2, \cdots, p_k\}$ be a basis of H_k. We have $t_p = \hat{U}(k) t_p \in C(\Omega)$, $p = 1, 2, \cdots, p_k$. We set $t'_p(\omega) = t_p(-\omega)$, $p = 1, 2, \cdots, p_k$. According to Lemma 6 [i], $\hat{U}(-k) t'_p(\omega) = \hat{U}(k) t_p(-\omega) = t_p(-\omega) = t'_p(\omega)$, which implies that $t'_p \in H_{-k}$, $p = 1, 2, \cdots, p_k$. The system of functions $\{t'_p | p = 1, 2, \cdots, p_k\}$ is linearly independent, so that $p_k \leq p_{-k}$ for every $k \in \mathfrak{N}$. Since $k \in \mathfrak{N}$ and $-k \in \mathfrak{N}$ at the same time, it follows that $p_{-k} \leq p_k$, $k \in \mathfrak{N}$, so that $p_k = p_{-k}$, $k \in \mathfrak{N}$.

Because $k \in \mathfrak{N}$ is real, the real and imaginary parts of the functions $t_p(\omega)$, $p = 1, 2, \cdots, p_k$, are themselves elements of H_k, and their linear shell coincides with H_k. It is clear from this that from the system of all real and imaginary parts of the functions t_p we can isolate a new basis of the subspace H_k whose elements are now real continuous functions. Making use of this possibility, we choose for every $k \in \mathfrak{N}$, $k > 0$, some basis $\{t_{kp} | p = 1, 2, \cdots, p_k\}$ of the subspace H_k consisting of real continuous functions. By Lemma 6 [iii], $((\omega n) f, f) < 0$ whatever the real function $f \in H_k$, $f \neq \vartheta$ ($k \in \mathfrak{N}$, $k > 0$). This enables us to subject the basis $\{t_{kp} | p = 1, 2, \cdots$ $\cdots, p_k\}$ to an ordinary orthogonalization procedure with the weight (ωn). Specifically, by induction over p we can construct a system of continuous real functions $\{v_{kp} | p = 1, 2, \cdots, p_k\}$ such that

a)
$$((\omega n), v_{kp}^2) < 0, \; p = 1, 2, \ldots, p_k,$$
$$((\omega n) v_{kp}, v_{kp'}) = 0, \; 1 \leqslant p < p' \leqslant p_k.$$

$$v_{k_1} = t_{k_1}.$$

b)
$$v_{kp} = t_{kp} - \sum_{s=1}^{p-1} \frac{((\omega n) v_{ks}, t_{kp})}{(\omega n, v_{ks}^2)} v_{ks}, \quad p = 2, \ldots, p_k.$$

c) v_{kp} is a linear combination of the vectors $t_{k1}, \cdots, t_{kp}$; $p = 1, 2, \cdots$ $\cdots, p_k$.

Let us set

$$\psi_{kp} = \frac{1}{\sqrt{-(\omega n, v_{kp}^2)}} v_{kp}, \; p = 1, 2, \ldots, p_k, \; k \in \mathfrak{N}, \; k > 0.$$

Clearly, $\{\psi_{kp} | p = 1, 2, \cdots, p_k\}$ is a basis of H_k, where $((\omega n) \psi_{kp}, \psi_{kp'}) = -\delta_{pp'} \operatorname{sgn} k$ for $1 \leq p \leq p' \leq p_k$; $k \in \mathfrak{N}$, $k > 0$.

The function ψ_{kp} for $k < 0$ can be defined on the basis of the ψ_{kp}, $k > 0$, already constructed.

Let $k \in \mathfrak{N}$, $k < 0$. We set $k_1 = -k$. Then $k_1 \in \mathfrak{N}$, $k > 0$, and, according to what we have proved, $p_k = p_{k_1}$. We define the functions ψ_{kp} by the relation

$$\psi_{kp}(\omega) = \psi_{k_1 p}(-\omega),\ \omega \in \Omega,\ p = 1, 2, \ldots, p_k.$$

Carrying out a similar construction for all $k \in \mathfrak{N}$, $k < 0$, we arrive at the system of functions

$$\Psi = \{\psi_{kp} \mid k \in \mathfrak{N},\ k \neq 0,\ p = 1, 2, \ldots, p_k\}.$$

It is clear that Ψ has properties [i] and [iii]. By Lemma 6 and the definition of ψ_{kp}, $k < 0$, we have

$$\psi_{kp} \in H_k,\ k \in \mathfrak{N},\ k < 0,\ p = 1, 2, \ldots, p_k.$$

It is clear that the functions $\{\psi_{kp} \mid p = 1, 2, \cdots, p_k\}$ are linearly independent. This means that they form a basis of H_k, $k \in \mathfrak{N}$, $k < 0$. It follows from this that the system Ψ has property [iv].

Finally, if $k \in \mathfrak{N}$, $k < 0$, $1 \leq p \leq p' \leq p_k$, then (compare Lemma 1)

$$((\omega \mathbf{n})\, \psi_{kp},\ \psi_{kp'}) = \int_{\Omega} (\omega \mathbf{n})\, \psi_{kp}(\omega)\, \psi_{kp'}(\omega)\, d\omega$$

$$= -\int_{\Omega} (-\omega \mathbf{n})\, \psi_{k_1 p}(-\omega)\, \psi_{k_1 p'}(-\omega)\, d\omega = -((\omega \mathbf{n})\, \psi_{k_1 p},\ \psi_{k_1 p'})$$

$$= \delta_{pp'} \operatorname{sgn} k_1 = -\delta_{pp'} \operatorname{sgn} k.$$

Thus Ψ also has property [ii]. The lemma is proved.

The system of functions whose existence is guaranteed by Lemma 15 is nonunique. To be specific, let us assume that some single system has been chosen from among these in such a way that

$$\psi_{\lambda_0 1} = |(\omega \mathbf{n},\ \Phi^2_{\lambda_0})|^{-1/2} \Phi_{\lambda_0} \tag{40}$$

(the possibility of such a choice is self-evident). We shall from now on denote this system by

$$\{\psi_{kp} \mid k \in \mathfrak{N},\ k \neq 0,\ p = 1, 2, \ldots, p_k\}.$$

Lemma 16. *Let G be an open set, $G \subset Z_0 \setminus \mathfrak{N}$, and let $\alpha(k)$ be a function with values in $L_2(\Omega)$ holomorphic in G. For every $k \in G$ there exists exactly one solution $\psi(k)$ of equation* (39), $\psi(k) = \hat{W}^{-1}(k)\, \alpha(k)$. *This solution $\psi(k)$ is a single-valued analytic function of the parameter k in G.*

Proof. Equation (39) is of the form

$$\hat{W}(k)\psi(k) = \alpha(k).$$

By Lemma 14, $\hat{W}(k)$ is invertible for $k \in G \subset Z_0 \setminus \mathfrak{N}$.

Hence $\psi(k) = \hat{W}^{-1}(k)\alpha(k)$ is the only solution of (39) in $L_2(\Omega)$.

Let us choose a $k_0 \in G$. Since $\alpha(k)$ is regular at the point $k = k_0$, there exists an $r \in (0, 1/r(k_0))$ $(r(k_0) = \|W^{-1}(k_0)\|)$ such that for $|k - k_0| < r$, $k \in G$, and $\alpha(k) = \sum_{\nu=0}^{\infty}(k - k_0)^\nu \alpha_\nu$ we have

$$\alpha_\nu \in L_2(\Omega),\ \|\alpha_\nu\| \leqslant Cr^{-\nu},\ \nu = 0, 1, 2, \ldots;\ C = \text{const} > 0.$$

We define the sequence $\{\psi_\nu \mid \nu = 0, 1, 2, \cdots\} \subset L_2(\Omega)$ by the relations

$$\psi_0 = \hat{W}^{-1}(k_0)\alpha_0,$$
$$\psi_\nu = \hat{W}^{-1}(k_0)[\alpha_\nu - (\omega\mathbf{n})\psi_{\nu-1}],\ \nu \geqslant 1.$$

Let us estimate $\|\psi_\nu\|$. Let $\eta_\nu = r^{-\nu}(k_0)\|\psi_\nu\|$, $\nu = 0, 1, 2, \cdots$. Then

$$\eta_0 \leqslant Cr(k_0),$$
$$\eta_\nu \leqslant \eta_{\nu-1} + \frac{Cr(k_0)}{(rr(k_0))^\nu},\quad \nu \geqslant 1.$$

Hence for $\nu \geq 1$ we have

$$\eta_\nu = \sum_{\nu'=1}^{\nu}(\eta_{\nu'} - \eta_{\nu'-1}) + \eta_0 \leqslant \sum_{\nu'=1}^{\nu}\frac{Cr(k_0)}{(rr(k_0))^{\nu'}} + Cr(k_0) \leqslant \frac{Cr(k_0)}{1 - rr(k_0)} \cdot \frac{1}{(rr(k_0))^\nu}.$$

Thus

$$\psi_\nu\| \leqslant \frac{Cr(k_0)}{1 - rr(k_0)} \cdot \frac{1}{r^\nu},\ \nu = 0, 1, 2, \ldots.$$

This estimate enables us to define the function

$$\chi(k) = \sum_{\nu=0}^{\infty}(k - k_0)^\nu \psi_\nu$$

with values in $L_2(\Omega)$ holomorphic in the disk $S(k_0 r) \equiv \{k \mid |k - k_0| < r\}$. For $k \in S(k_0, r)$ we have

$$\hat{W}(k)\chi(k) = \hat{W}(k_0)\chi(k) + (k - k_0)(\omega\mathbf{n})\chi(k)$$
$$= \sum_{\nu=0}^{\infty}(k - k_0)^\nu \hat{W}(k_0)\psi_\nu + (k - k_0)(\omega\mathbf{n})\chi(k)$$
$$= \alpha_0 + \sum_{\nu=1}^{\infty}(k - k_0)^\nu[\alpha_\nu - (\omega\mathbf{n})\psi_{\nu-1}] + (k - k_0)(\omega\mathbf{n})\chi(k)$$
$$= \alpha(k) - (\omega\mathbf{n})\sum_{\nu=1}^{\infty}(k - k_0)^\nu\psi_{\nu-1} + (k - k_0)(\omega\mathbf{n})\chi(k) = \alpha(k).$$

It follows from this that

$$\psi(\mathrm{k}) = \hat{W}^{-1}(\mathrm{k})\,\alpha(\mathrm{k}) = \chi(\mathrm{k})$$

for $\mathrm{k} \in S(\mathrm{k}_0, r)$. The lemma is proved.

Lemma 17. *Let* $\mathrm{k} \in \mathfrak{N}$, $\mathrm{k} \neq 0$, *and let* $\{\alpha_\nu \mid \nu = 0, 1, 2, \cdots\}$ *be an arbitrary sequence of vectors from* $L_2(\Omega)$. *There exists exactly one sequence* $\{\psi_\nu \mid \nu = -1, 0, 1, 2, \cdots\} \subset L_2(\Omega)$ *such that*

$$(1 + \mathrm{k}\omega\mathbf{n})\,\psi_{-1} = \hat{g}\psi_{-1},$$

$$(1 + \mathrm{k}\omega\mathbf{n})\,\psi_\nu = \hat{g}\psi_\nu + \alpha_\nu - (\omega\mathbf{n})\,\psi_{\nu-1}, \quad \nu \geqslant 0. \tag{41}$$

This sequence has the following properties:

$$\psi_{-1} = -(\operatorname{sgn}\mathrm{k}) \sum_{p=1}^{p_\mathrm{k}} (\alpha_0, \psi_{\mathrm{k}p})\,\psi_{\mathrm{k}p}; \tag{42}$$

$$\alpha_\nu - (\omega\mathbf{n})\,\psi_{\nu-1} \in \mathfrak{L}_\mathrm{k}, \quad \nu \geqslant 0;$$

$$\psi_\nu = \hat{N}_\mathrm{k}^{-1}(\alpha_\nu - (\omega\mathbf{n})\,\psi_{\nu-1}) + \sum_{p=1}^{p_\mathrm{k}} B_{\nu p}\psi_{\mathrm{k}p}, \quad \nu \geqslant 0,$$

where the numbers $B_{\nu p}$ *are given by the relation*

$$B_{\nu p} = -(\operatorname{sgn}\mathrm{k})(\alpha_{\nu+1}, \psi_{\mathrm{k}p}) + (\operatorname{sgn}\mathrm{k})(\hat{N}_\mathrm{k}^{-1}(\alpha_\nu - (\omega\mathbf{n})\,\psi_{\nu-1}), (\omega\mathbf{n})\,\psi_{\mathrm{k}p}),$$
$$\nu \geqslant 0, \; p = 1, 2, \ldots, p_\mathrm{k}.$$

(*The subspace* $\mathfrak{L}_\mathrm{k}$ *and the operator* $\hat{N}_\mathrm{k}$ *are defined in Lemma* 14).

Proof. The expressions for ψ_ν given in the formulation of the lemma enable us to construct the sequence $\{\psi_\nu \mid \nu = -1, 0, 1, 2, \cdots\}$ by induction over ν. To do this we define ψ_{-1} in accordance with formula (42) and assume that for some $s \geq -1$ we have already constructed a finite sequence $\{\psi_\nu \mid -1 \leq \nu \leq s\} \subset L_2(\Omega)$ such that for $0 \leq \nu \leq s$ (if such ν exist) the ψ_ν satisfy the relations

$$\alpha_\nu - (\omega\mathbf{n})\,\psi_{\nu-1} \in \mathfrak{L}_\mathrm{k},$$

$$\psi_\nu = \hat{N}_\mathrm{k}^{-1}(\alpha_\nu - (\omega\mathbf{n})\,\psi_{\nu-1}) + \sum_{p=1}^{p_\mathrm{k}} [-\operatorname{sgn}\mathrm{k}\cdot(\alpha_{\nu+1}, \psi_{\mathrm{k}p})$$
$$+ \operatorname{sgn}\mathrm{k}\cdot(\hat{N}_\mathrm{k}^{-1}(\alpha_\nu - (\omega\mathbf{n})\,\psi_{\nu-1}), (\omega\mathbf{n})\,\psi_{\mathrm{k}p})]\,\psi_{\mathrm{k}p}. \tag{43}$$

We note that $\alpha_{s+1} - (\omega\mathbf{n})\psi_s \in \mathfrak{L}_\mathrm{k}$. In fact, if $s = -1$, then for every p, $1 \leq p \leq p_\mathrm{k}$, we have

$$(\alpha_{s+1} - (\omega \mathbf{n})\psi_s, \psi_{kp}) = \Big(\alpha_0 - (\omega \mathbf{n})\Big(-\operatorname{sgn} k \cdot \sum_{p'=1}^{p_k} (\alpha_0, \psi_{kp'})\psi_{kp'}\Big), \psi_{kp}\Big)$$

$$= (\alpha_0, \psi_{kp}) + \operatorname{sgn} k \sum_{p'=1}^{p_k} (\alpha_0, \psi_{kp'})((\omega \mathbf{n})\psi_{kp'}, \psi_{kp})$$

$$= (\alpha_0, \psi_{kp}) - \sum_{p'=1}^{p_k} (\alpha_0, \psi_{kp'})\delta_{pp'} = 0$$

(compare Lemma 15). On the other hand, if $s \geq 0$, then

$$(\alpha_{s+1} - (\omega \mathbf{n})\psi_s, \psi_{kp}) = (\alpha_{s+1}, \psi_{kp}) - (\hat{N}_k^{-1}(\alpha_s - (\omega \mathbf{n})\psi_{s-1}), (\omega \mathbf{n})\psi_{kp})$$

$$- \sum_{p'=1}^{p_k} [-\operatorname{sgn} k(\alpha_{s+1}, \psi_{kp'}) + \operatorname{sgn} k(\hat{N}_k^{-1}(\alpha_s - (\omega \mathbf{n})\psi_{s-1}), (\omega \mathbf{n})\psi_{kp'})]$$

$$\times (\psi_{kp'}, (\omega \mathbf{n})\psi_{kp}) = (\alpha_{s+1}, \psi_{kp}) - (\hat{N}_k^{-1}(\alpha_s - (\omega \mathbf{n})\psi_{s-1}), (\omega \mathbf{n})\psi_{kp})$$

$$- [(\alpha_{s+1}, \psi_{kp}) - (\hat{N}_k^{-1}(\alpha_s - (\omega \mathbf{n})\psi_{s-1}), (\omega \mathbf{n})\psi_{kp})] = 0, \ 1 \leqslant p \leqslant p_k.$$

The inclusion $\alpha_{s+1} - (\omega \mathbf{n})\psi_s \in \mathfrak{L}_k$ enables us to define the vector ψ_{s+1} as follows:

$$\psi_{s+1} = \hat{N}_k^{-1}(\alpha_{s+1} - (\omega \mathbf{n})\psi_s)$$

$$+ \sum_{p=1}^{p_k} [-\operatorname{sgn} k \cdot (\alpha_{s+2}, \psi_{kp}) + \operatorname{sgn} k \cdot (\hat{N}_k^{-1}(\alpha_{s+1} - (\omega \mathbf{n})\psi_s), (\omega \mathbf{n})\psi_{kp})]\psi_{kp}.$$

Let us incorporate ψ_{s+1} in the initial sequence $\{\psi_\nu | -1 \leq \nu \leq s\}$. This yields the new initial sequence $\{\psi_\nu | -1 \leq \nu \leq s+1\}$ in which ψ_{-1} has its old form. Relations (43), $0 \leq \nu \leq s+1$, remain valid for this new sequence. In fact, the inclusion $\alpha_\nu - (\omega \mathbf{n})\psi_{-1} \in \mathfrak{L}_k$ for $0 \leq \nu \leq s$ is fulfilled by virtue of the inductive assumption; it is proved above for $\nu = s+1$. The second relation of (43) for $0 \leq \nu \leq s$ also follows from the inductive assumption; for $\nu = s+1$ it follows from the definition of ψ_{s+1}.

Thus the new sequence has the same properties as the initial sequence. This enables us to replace s by $s+1$ and to repeat the above arguments. Let us construct the vectors $\psi_{s+2}, \cdots$. This gives rise to the infinite sequence $\{\psi_\nu | \nu = -1, 0, 1, 2, \cdots\}$ whose elements are subject to conditions (43) for $\nu \geq 0$; ψ_{-1} is defined by formula (42). We shall now show that this sequence satisfies system (41).

We have

$$\hat{W}(k)\psi_{-1} = -(\operatorname{sgn} k) \sum_{p=1}^{p_k} (\alpha_0, \psi_{kp})\hat{W}(k)\psi_{kp} = \vartheta,$$

or, which is the same thing,

$$(1 + k\omega n)\,\psi_{-1} = \hat{g}\psi_{-1}.$$

For $\nu \geq 0$ (compare (43)) we have

$$\hat{W}(k)\,\psi_\nu = \hat{W}(k)\,\hat{N}_k^{-1}(\alpha_\nu - (\omega n)\,\psi_{\nu-1}) + \sum_{p=1}^{p_k} [\ldots]\, \hat{W}(k)\,\psi_{kp} = \alpha_\nu - (\omega n)\,\psi_{\nu-1},$$

so that

$$(1 + k\omega n)\,\psi_\nu = \hat{g}\psi_\nu + \alpha_\nu - (\omega n)\,\psi_{\nu-1},\ \nu \geqslant 0.$$

We see that the sequence $\{\psi_\nu | \nu = -1, 0, 1, 2, \cdots\}$ has all of the properties given in the lemma. It remains for us to prove that if some sequence $\{\psi'_\nu | \nu = -1, 0, 1, 2, \cdots\}$ satisfies system (41), then $\psi'_\nu = \psi_\nu$, $\nu = -1, 0, 1, 2, \cdots$.

Let us assume that $\{\psi'_\nu\} \neq \{\psi_\nu\}$. Let ν_0 be such that $\psi'_{\nu_0} \neq \psi_{\nu_0}$, but $\psi'_\nu = \psi_\nu$ for $-1 \leq \nu \leq \nu_0 - 1$. We set $\delta = \psi'_{\nu_0} - \psi_{\nu_0}$, $\delta \neq \vartheta$. If $\nu_0 = -1$, then, clearly, $(1 + k\omega n)\delta = \hat{g}\delta$. But if $\nu_0 > -1$, then

$$(1 + k\omega n)\,\psi'_{\nu_0} = \hat{g}\psi'_{\nu_0} + \alpha_{\nu_0} - (\omega n)\,\psi'_{\nu_0-1},$$
$$(1 + k\omega n)\,\psi_{\nu_0} = \hat{g}\psi_{\nu_0} + \alpha_{\nu_0} - (\omega n)\,\psi_{\nu_0-1}.$$

Subtracting the second relation from the first and recalling that $\psi'_{\nu_0-1} = \psi_{\nu_0-1}$, we obtain $(1 + k\omega n)\delta = \hat{g}\delta$. This means that, in any case, $\hat{W}(k)\delta = \vartheta$. From this it follows that $\delta \in H_k$ and $\delta = \sum_{p=1}^{p_k} a_p \psi_{kp} \neq \vartheta$.

Let $\delta_1 = \psi'_{\nu_0+1} - \psi_{\nu_0+1}$. Then, clearly,

$$(1 + k\omega n)\,\delta_1 = \hat{g}\delta_1 - (\omega n)\,\delta.$$

Let us choose a natural number $\bar{p}$, $1 \leq \bar{p} \leq p_k$, and take the scalar product of the right side of the latter expression and $\psi_{k\bar{p}}$:

$$((1 + k\omega n)\,\delta_1,\ \psi_{k\bar{p}}) = (\hat{g}\delta_1,\ \psi_{k\bar{p}}) - \sum_{p=1}^{p_k} a_p\,((\omega n)\,\psi_{kp},\ \psi_{k\bar{p}})$$
$$= (\delta_1,\ \hat{g}\psi_{k\bar{p}}) + a_{\bar{p}} \operatorname{sgn} k = a_{\bar{p}} \operatorname{sgn} k + (\delta_1,\ (1 + k\omega n)\,\psi_{k\bar{p}}).$$

This means that $a_{\bar{p}} \operatorname{sgn} k = 0$, $a_{\bar{p}} = 0$. The arbitrariness of $\bar{p}$ implies that $\sum_{p=1}^{p_k} |a_p| = 0$, which contradicts the condition $\delta \neq \vartheta$. This contradiction means that $\{\psi'_\nu\} = \{\psi_\nu\}$. The lemma is proved.

Lemma 18. *Let* $g_0 = 1$. *Then* $0 \in \mathfrak{N}$ *and* $\mathfrak{L}_0 = \{f \,|\, f \in L_2(\Omega),\ (f, 1) = 0\}$. *Consider an arbitrary sequence*

$$\{\alpha_\nu \,|\, \nu = -1, 0, 1, 2, \ldots\} \subset L_2(\Omega),\ \alpha_{-1} = \vartheta.$$

There exists exactly one sequence

$$\{\psi_\nu \mid \nu = -2, -1, 0, 1, 2, 3, \ldots\} \subset L_2(\Omega)$$

such that

$$\psi_{-2} = \hat{g}\psi_{-2},$$

$$\psi_\nu = \hat{g}\psi_\nu + \alpha_\nu - (\omega\mathbf{n})\,\psi_{\nu-1}, \quad \nu \geqslant -1. \tag{44}$$

This sequence has the following properties:

$$\psi_{-2}(\omega) = -\frac{3}{4\pi}(1-g_1)(\alpha_0, 1), \quad \omega \in \Omega, \tag{45}$$

$$\alpha_\nu - (\omega\mathbf{n})\,\psi_{\nu-1} \in \mathcal{L}_0, \ \nu \geqslant -1, \tag{46}$$

$$\psi_\nu = \hat{N}_0^{-1}(\alpha_\nu - (\omega\mathbf{n})\,\psi_{\nu-1}) + B_\nu, \quad \nu \geqslant -1, \tag{47}$$

where the number B_ν *is defined by the relation*

$$B_\nu = -\frac{3}{4\pi}(1-g_1)(\alpha_{\nu+2}, 1) + \frac{3}{4\pi}(\alpha_{\nu+1}, \omega\mathbf{n}) - \frac{3}{4\pi}\frac{(\alpha_\nu, (\omega\mathbf{n})^2)}{1-g_2} + \frac{3}{4\pi}\frac{(\psi_{\nu-1}, (\omega\mathbf{n})^3)}{1-g_2}, \quad \nu \geqslant -1. \tag{48}$$

Proof. We have $0 = \lambda_0 \in \mathfrak{N}$ for $g_0 = 1$ (Theorem 10). As already noted, the subspace $H_0 = H_{\lambda_0}$ is one-dimensional, and $\Phi_0 \in H_0$. But $\Phi_0(\omega) \equiv 1$ (Theorem 11), which implies that $\mathfrak{L}_0 = \{f \mid f \in L_2(\Omega), (f, 1) = 0\}$.

Let us choose an arbitrary orthonormal basis σ of the space R_3 with the third unit vector $\mathbf{n}$. By Lemma 2,

$$Y_{\sigma n}^{(0)}(\omega) = \left(\frac{2n+1}{4\pi}\right)^{1/2} P_n(\omega\mathbf{n}), \quad n = 0, 1, 2, \ldots.$$

Because spherical functions are orthogonal, $Y_{\sigma_n}^{(m)} \in \mathfrak{L}_0$ for $n > 0$, $-n \leq m \leq n$. In particular,

$$\omega\mathbf{n} = P_1(\omega\mathbf{n}) = \sqrt{\frac{4\pi}{3}}\, Y_{\sigma 1}^{(0)} \in \mathcal{L}_0;$$

$$(\omega\mathbf{n})^2 - \frac{1}{3} = \frac{2}{3} P_2(\omega\mathbf{n}) = \frac{2}{3}\sqrt{\frac{4\pi}{5}}\, Y_{\sigma 2}^{(0)} \in \mathcal{L}_0.$$

We note that

$$\hat{N}_0(\omega\mathbf{n}) = \hat{W}(0)(\omega\mathbf{n}) = \omega\mathbf{n} - \hat{g}(\omega\mathbf{n}) = (1-g_1)(\omega\mathbf{n}),$$

$$\hat{N}_0\left[(\omega\mathbf{n})^2 - \frac{1}{3}\right] = (1-g_2)\left[(\omega\mathbf{n})^2 - \frac{1}{3}\right].$$

Hence

$$\hat{N}_0^{-1}(\omega\mathbf{n}) = \frac{1}{1-g_1}\,\omega\mathbf{n},$$
$$\hat{N}_0^{-1}\left[(\omega\mathbf{n})^2 - \frac{1}{3}\right] = \frac{1}{1-g_2}\left[(\omega\mathbf{n})^2 - \frac{1}{3}\right]. \tag{49}$$

Let $\{\alpha_\nu \mid \nu \geq -1\} \subset L_2(\Omega)$, $\alpha_{-1} = \vartheta$. We define ψ_{-2} in accordance with (45) and assume that the finite sequence $\{\psi_\nu \mid -2 \leq \nu < s\}$, $s \geq -2$, whose elements satisfy relations (46)–(48) for $-1 \leq \nu \leq s$ (if $s \geq -1$) has already been constructed. As in proving Lemma 17, we shall attempt to choose ψ_{s+1} in such a way that these relations remain valid for $\nu = s+1$ We note first of all that

$$\alpha_{s+1} - (\omega\mathbf{n})\,\psi_s \in \mathscr{L}_0. \tag{50}$$

In fact, if $s = -2$, then

$$(\alpha_{s+1} - (\omega\mathbf{n})\,\psi_s,\, 1) = -(\psi_{-2},\, \omega\mathbf{n}) = \frac{3}{4\pi}(1-g_1)(\alpha_0,\, 1)(1,\, \omega\mathbf{n}) = 0.$$

If $s = -1$, then

$$(\alpha_{s+1} - (\omega\mathbf{n})\,\psi_s,\, 1) = (\alpha_0,\, 1) - \left(\hat{N}_0^{-1}\left(\frac{3}{4\pi}(1-g_1)(\alpha_0,\, 1)\,\omega\mathbf{n}\right) + B_1,\, \omega\mathbf{n}\right)$$
$$= (\alpha_0,\, 1) - \frac{3}{4\pi}(\alpha_0,\, 1)(1,\, (\omega\mathbf{n})^2) = 0.$$

Finally, for $s \geq 0$ we have

$$(\alpha_{s+1} - (\omega\mathbf{n})\,\psi_s,\, 1) = (\alpha_{s+1},\, 1) - (\psi_s,\, \omega\mathbf{n}) = (\alpha_{s+1},\, 1)$$
$$- (\alpha_s - (\omega\mathbf{n})\,\psi_{s-1},\, \hat{N}_0^{-1}(\omega\mathbf{n})) - (B_s,\, \omega\mathbf{n}) = (\alpha_{s+1},\, 1) - \frac{1}{1-g_1}(\alpha_s,\, \omega\mathbf{n})$$
$$+ \frac{1}{1-g_1}(B_{s-1},\, (\omega\mathbf{n})^2) + \frac{1}{1-g_1}(\hat{N}_0^{-1}(\alpha_{s-1} - (\omega\mathbf{n})\,\psi_{s-2}),\, (\omega\mathbf{n})^2).$$

We recall that $\hat{N}_0^{-1}(\alpha_{s-1} - (\omega\mathbf{n})\psi_{s-2}) \in \mathscr{L}_0$, so that

$$\left(\hat{N}_0^{-1}(\alpha_{s-1} - (\omega\mathbf{n})\,\psi_{s-2}),\, \frac{1}{3}\right) = 0.$$

Hence

$$(\alpha_{s+1} - (\omega\mathbf{n})\,\psi_s,\, 1) = (\alpha_{s+1},\, 1) - \frac{(\alpha_s,\, \omega\mathbf{n})}{1-g_1} + \frac{4\pi}{3}\frac{B_{s-1}}{1-g_1}$$

$$+ \frac{1}{1-g_1}\left(\hat{N}_0^{-1}(\alpha_{s-1} - (\omega\mathbf{n})\,\psi_{s-2}),\, (\omega\mathbf{n})^2 - \frac{1}{3}\right) = (\alpha_{s+1},\, 1) - \frac{(\alpha_s,\, \omega\mathbf{n})}{1-g_1}$$
$$+ \frac{4\pi}{3}\frac{B_{s-1}}{1-g_1} + \frac{1}{1-g_1}\cdot\frac{1}{1-g_2}(\alpha_{s-1} - (\omega\mathbf{n})\,\psi_{s-2},\, (\omega\mathbf{n})^2) =$$

$$= (\alpha_{s+1}, 1) - \frac{(\alpha_s, \omega n)}{1-g_1} + \frac{(\alpha_{s-1}, (\omega n)^2)}{(1-g_1)(1-g_2)} - \frac{(\psi_{s-2}, (\omega n)^3)}{(1-g_1)(1-g_2)}$$

$$+ \frac{4\pi}{3}\frac{1}{1-g_1}\left[-\frac{3}{4\pi}(1-g_1)(\alpha_{s+1}, 1) + \frac{3}{4\pi}(\alpha_s, \omega n) - \frac{3}{4\pi}\frac{(\alpha_{s-1}, (\omega n)^2)}{1-g_2}\right.$$

$$\left. + \frac{3}{4\pi}\frac{(\psi_{s-2}, (\omega n)^3)}{1-g_2}\right] = 0.$$

Inclusion (50) has been proved. This enables us to define ψ_{s+1} in accordance with formulas (47) and (48):

$$\psi_{s+1} = \hat{N}_0^{-1}(\alpha_{s+1} - (\omega n)\psi_s) + B_{s+1},$$

$$B_{s+1} = -\frac{3}{4\pi}(1-g_1)(\alpha_{s+3}, 1) + \frac{3}{4\pi}(\alpha_{s+2}, \omega n)$$

$$- \frac{3}{4\pi}\frac{(\alpha_{s+1}, (\omega n)^2)}{1-g_2} + \frac{3}{4\pi}\frac{(\psi_s, (\omega n)^3)}{1-g_2}.$$

Let us incorporate ψ_{s+1} in the initial sequence $\{\psi'_\nu \mid -2 \le \nu \le s\}$. As is readily verifiable, the elements of the new sequence $\{\psi_\nu \mid -2 \le \nu \le s+1\}$ satisfy relations (45) and (46)–(48) for $-1 \le \nu \le s+1$. Replacing s by $s+1$ and repeating the argument, we obtain the vector ψ_{s+2}, etc.

We arrive at the infinite sequence $\{\psi_\nu \mid \nu \ge -2\} \subset L_2(\Omega)$ for which conditions (45–(48) are fulfilled. We can show that it is a solution of system (44).

Since $\psi_{-2}(\omega) \equiv \text{const}$, we see that

$$\hat{g}\psi_{-2} = \psi_{-2}.$$

For $\nu \ge -1$ we have

$$\hat{W}(0)\psi_\nu = \hat{W}(0)\hat{N}_0^{-1}(\alpha_\nu - (\omega n)\psi_{\nu-1}) + \hat{W}(0)B_\nu$$
$$= \alpha_\nu - (\omega n)\psi_{\nu-1} + (B_\nu - \hat{g}B_\nu) = \alpha_\nu - (\omega n)\psi_{\nu-1},$$

so that

$$\psi_\nu = \hat{g}\psi_\nu + \alpha_\nu - (\omega n)\psi_{\nu-1}, \quad \nu \geqslant -1.$$

The resulting sequence $\{\psi_\nu \mid \nu \ge -2\}$ satisfies all the requirements of the lemma.

It remains for us to prove that system (44) admits of not more than one solution. Let $\{\psi'_\nu \mid \nu \ge -2\} \subset L_2(\Omega)$ be a solution of (44). Let us suppose that $\{\psi'_\nu\} \ne \{\psi_\nu\}$. There then exists a $\nu_0 \ge -2$ such that $\psi'_{\nu_0} \ne \psi_{\nu_0}$, while for $-2 \le \nu \le \nu_0 - 1$ (if such ν exist) we have $\psi'_\nu = \psi_\nu$. Let $\delta = \psi_{\nu_0} - \psi_{\nu_0}$, $\delta \ne \vartheta$. By the choice of ν_0, (44) implies that $\hat{W}(0)\delta = \vartheta$, $\delta \in H_0$, and $\delta(\omega) \equiv C = \text{const} \ne 0$, $\omega \in \Omega$. We note that

$$\psi'_{\nu_0+1} = \hat{g}\psi'_{\nu_0+1} + \alpha_{\nu_0+1} - (\omega n)\psi'_{\nu_0},$$
$$\psi_{\nu_0+1} = \hat{g}\psi_{\nu_0+1} + \alpha_{\nu_0+1} - (\omega n)\psi_{\nu_0}.$$

We set $\delta_1 = \psi'_{\nu_0+1} - \psi_{\nu_0+1}$. Clearly,

$$\delta_1(\omega) = \hat{g}\delta_1(\omega) - C\omega\mathbf{n}, \quad \omega \in \Omega.$$

Let $\beta(\omega) = -C\omega\mathbf{n}/(1-g_1)$, $\omega \in \Omega$. Then $\beta(\omega) - \hat{g}\beta(\omega) = -C\omega\mathbf{n}$, so that $(\delta_1 - \beta) = \hat{g}(\delta_1 - \beta)$. Thus $\delta_1(\omega) - \beta(\omega) = C_1 = \text{const}$, $\omega \in \Omega$.

Finally, let us construct the difference $\delta_2 = \psi'_{\nu_0+2} - \psi_{\nu_0+2}$. Equating formulas (44) for ψ'_{ν_0+2} and ψ_{ν_0+2}, we obtain

$$\delta_2 - \hat{g}\delta_2 = -(\omega\mathbf{n})\,\delta_1 = -(\omega\mathbf{n})\,(\beta + C_1).$$

Let us take the scalar product of this equation and unity,

$$-(\beta, \omega\mathbf{n}) - C_1(\omega\mathbf{n}, 1) = (\delta_2, 1) - (\hat{g}\delta_2, 1) = (\delta_2, 1) - (\delta_2, \hat{g}1) = 0.$$

Since $(\omega\mathbf{n}, 1) = 0$, we obtain

$$\frac{C}{1-g_1}(1, (\omega\mathbf{n})^2) = 0.$$

But $C \neq 0$, so that $(1, (\omega\mathbf{n})^2) = 0$, which is not true. This contradiction means that $\{\psi'_\nu\} = \{\psi_\nu\}$. The lemma is proved.

In addition to ψ_{-2}, we shall also need the explicit form of ψ_{-1}. The formulas of Lemma 18 enable us to find it. We have

$$\psi_{-1} = \hat{N}_0^{-1}(\alpha_{-1} - (\omega\mathbf{n})\,\psi_{-2}) + B_{-1} = \frac{3}{4\pi}(\alpha_0, 1)\,\omega\mathbf{n} + B_{-1},$$

$$B_{-1} = -\frac{3}{4\pi}(1-g_1)(\alpha_1, 1) + \frac{3}{4\pi}(\alpha_0, \omega\mathbf{n}) - \frac{3}{4\pi}\,\frac{(\alpha_{-1}, (\omega\mathbf{n})^2)}{1-g_2} + \frac{3}{4\pi}\,\frac{(\psi_{-2}, (\omega\mathbf{n})^3)}{1-g_2}.$$

But $\alpha_{-1} = \vartheta$, and $\psi_{-2}(\omega) = \text{const}$, so that $(\psi_{-2}, (\omega\mathbf{n})^3) = 0$, and we find that

$$B_{-1} = -\frac{3}{4\pi}(1-g_1)(\alpha_1, 1) + \frac{3}{4\pi}(\alpha_0, \omega\mathbf{n}).$$

Thus, under the conditions of Lemma 18, we find that

$$\psi_{-1}(\omega) = \frac{3}{4\pi}(\alpha_0, 1)\,\omega\mathbf{n} - \frac{3}{4\pi}(1-g_1)(\alpha_1, 1) + \frac{3}{4\pi}(\alpha_0, \omega\mathbf{n}). \tag{51}$$

for $\omega \in \Omega$.

Theorem 16. *Let G be an open set, $G \in Z_0$, and let $\alpha(\mathbf{k})$ be an abstract function with values in $L_2(\Omega)$ which is holomorphic in G. For every $\mathbf{k} \in G\backslash\mathfrak{N}$ the space $L_2(\Omega)$ contains exactly one solution $\psi(\mathbf{k})$ of equation* (39), *and $\psi(\mathbf{k}) = \hat{W}^{-1}(\mathbf{k})\,\alpha(\mathbf{k})$. This solution $\psi(\mathbf{k})$ is a single-valued analytic function of $\mathbf{k} \in G\backslash\mathfrak{N}$.*

Let $\mathbf{k}_0 \in G \cap \mathfrak{N}$. There exists an $r > 0$ for which $S(\mathbf{k}_0, r) \equiv \{\mathbf{k} \mid |\mathbf{k} - \mathbf{k}_0| < r\} \subset G$, $S(\mathbf{k}_0, r) \cap \mathfrak{N} = \{\mathbf{k}_0\}$. Choose such an r. It turns out that

[i] *If* $g_0 = 1$ *and* $k_0 = 0$, *then*

$$\psi(k, \omega) = -\frac{3}{4\pi}\frac{1-g_1}{k^2}(\alpha(0), 1)$$

$$+\frac{3}{4\pi k}[(\alpha(0), 1)\omega n - (1-g_1)(\alpha'(0), 1) + (\alpha(0), \omega n)] + u(k,\omega).$$

for $k \in S(0, r)$, $k \neq 0$, $\omega \in \Omega$; *here* $(\alpha'(0) = d\alpha(k)/dk|_{k=0})$.

[ii] *If* $k_0 \neq 0$, *then*

$$\psi(k, \omega) = -\frac{\operatorname{sgn} k_0}{k - k_0}\sum_{p=1}^{p_{k_0}}(\alpha(k_0), \psi_{k_0 p})\psi_{k_0 p}(\omega) + u(k, \omega)$$

for $k \in S(k_0, r)$, $k \neq k_0$, $\omega \in \Omega$.

In both cases $u(k)$ *is a function of* k *with values in* $L_2(\Omega)$ *which is holomorphic in* $S(k_0, r)$.

Proof. The single-valuedness and analyticity in the domain $G_1 = G\backslash\mathfrak{N}$ of the solution $\psi(k) = \hat{W}^{-1}(k)\alpha(k)$ of equation (16) are established in Lemma 16.

By Theorem 5, to every $k_0 \in \mathfrak{N} \cap G$ we can assign an $r = r_{k_0} > 0$ such that $S(k_0, r) \subset G$, $S(k_0, r) \cap \mathfrak{N} = \{k_0\}$. Let us choose a $k_0 \in G \cap \mathfrak{N}$ and an $r = r_{k_0}$.

First let us consider the case $k_0 \neq 0$. We set

$$B = p_{k_0}\left(\max_{1\leq p\leq p_{k_0}}\|\psi_{k_0 p}\|\right)^2, \quad A = (1+B)T(k_0).$$

($T(k_0) = \|\hat{N}_{k_0}^{-1}\|$. Compare Lemma 14.) There exists an $r' > 0$ such that

$$\alpha(k) = \sum_{\nu=0}^{\infty}(k-k_0)^{\nu}\alpha_{\nu},$$

$$\alpha_{\nu} \in L_2(\Omega), \quad \|\alpha_{\nu}\| \leq C(r')^{-\nu}, \quad \nu = 0, 1, 2, \ldots; C = \text{const} > 0$$

for $|k - k_0| < r'$, $k \in G$.

Now let us choose a number r_1 such that $0 < r_1 < \min\{r, r', A^{-1}\}$. Let us assign a sequence $\{\psi_{\nu}|\nu \geq -1\} \subset L_2(\Omega)$ satisfying the requirements of Lemma 17 to the sequence of coefficients $\{\alpha_{\nu}|\nu = 0, 1, 2, \cdots\}$, and let us estimate $\|\psi_{\nu}\|$. We have

$$\|\psi_{-1}\| \leq \sum_{p=1}^{p_{k_0}}\|\alpha_0\|\,\|\psi_{k_0 p}\|^2 \leq B\|\alpha_0\|.$$

For $\nu \geq 0$ we find that

$$\|\psi_{\nu}\| \leq T(k_0)(\|\alpha_{\nu}\| + \|\psi_{\nu-1}\|) + \left(\max_{1\leq p\leq p_{k_0}}\|\psi_{k_0 p}\|\right)\sum_{p=1}^{p_{k_0}}|B_{\nu p}|,$$

$$|B_{\nu p}| \leq \|\alpha_{\nu+1}\|\,\|\psi_{k_0 p}\| + T(k_0)(\|\alpha_{\nu}\| + \|\psi_{\nu-1}\|)\|\psi_{k_0 p}\|.$$

This means that

$$\|\psi_\nu\| \leq T(k_0)(\|\alpha_\nu\| + \|\psi_{\nu-1}\|) + B(\|\alpha_{\nu+1}\| + T(k_0)(\|\alpha_\nu\| + \|\psi_{\nu-1}\|))$$
$$= B\|\alpha_{\nu+1}\| + A(\|\alpha_\nu\| + \|\psi_{\nu-1}\|) \leq A\|\psi_{\nu-1}\| + A\frac{C}{r_1^\nu} + B\frac{C}{r_1^{\nu+1}}$$
$$= A\|\psi_{\nu-1}\| + \frac{D}{r_1^\nu}, \quad \nu \geq 0.$$

Here $D = C(A + Br_1^{-1})$.

Let us set $\xi_\nu = A^{-\nu}\|\psi_\nu\|$ for $\nu \geq -1$. Then

$$\xi_{-1} \leq AB\|\alpha_0\| \leq ABC \leq \frac{BC}{r_1} \leq D,$$

and for $\nu \geq 0$ we have

$$\xi_\nu \leq \xi_{\nu-1} + \frac{D}{(Ar_1)^\nu}.$$

Hence

$$\xi_\nu = \sum_{\nu'=0}^{\nu} (\xi_{\nu'} - \xi_{\nu'-1}) + \xi_{-1} \leq \sum_{\nu'=0}^{\nu} \frac{D}{(Ar_1)^{\nu'}} + D \leq \frac{2D}{1 - Ar_1} \cdot \frac{1}{(Ar_1)^\nu}, \quad \nu \geq 0.$$

This means that

$$\|\psi_\nu\| \leq \frac{2D}{1 - Ar_1} \cdot \frac{1}{r_1^\nu}, \quad \nu \geq 0.$$

We set

$$\psi^{(1)}(k) = \sum_{\nu=-1}^{\infty} (k - k_0)^\nu \psi_\nu$$

for $k \in S(k_0, r_1)$, $k \neq 0$. The convergence of this series in the metric of $L_2(\Omega)$ is ensured by the above estimate.

Let us show that $\psi^{(1)}(k)$ satisfies equation (39).

Let $0 < |k - k_0| < r_1 < r'$. Then

$$(1 + k\omega n)\psi^{(1)}(k) = (1 + k_0\omega n)\psi^{(1)}(k) + (k - k_0)(\omega n)\psi^{(1)}(k)$$
$$= (1 + k_0\omega n)\sum_{\nu=-1}^{\infty}(k - k_0)^\nu\psi_\nu + (\omega n)\sum_{\nu=0}^{\infty}(k - k_0)^\nu\psi_{\nu-1}$$
$$= (k - k_0)^{-1}(1 + k_0\omega n)\psi_{-1} + \sum_{\nu=0}^{\infty}(k - k_0)^\nu[(1 + k_0\omega n)\psi_\nu + (\omega n)\psi_{\nu-1}].$$

Now let us make use of equations (41). We have

$$(1 + k\omega n)\psi^{(1)}(k) = (k - k_0)^{-1}\hat{g}\psi_{-1} + \sum_{\nu=0}^{\infty}(k - k_0)^\nu(\hat{g}\psi_\nu + \alpha_\nu) = \hat{g}\psi^{(1)}(k) + \alpha(k).$$

Owing to the choice of r_1, the condition $0 \leq |k - k_0| < r_1$ implies that $k \in G \setminus \mathfrak{N}$. But equation (39) admits of only the one solution $\psi(k)$ for such values of k.

Hence for $k \in S(k_0, r_1)$, $k \neq k_0$, we have

$$\psi(k) = \psi^{(1)}(k) = (k - k_0)^{-1}\psi_{-1} + u(k),$$

$u(k) = \Sigma_{\nu=0}^{\infty}(k - k_0)^{\nu}\psi_{\nu}$ is a function holomorphic in $S(k_0, r_1)$, and $\psi(k)$ and $(k - k_0)^{-1}\psi_{-1}$ are holomorphic with respect to k in $S(k_0, r)\backslash\{k_0\}$ (since, by the choice of $r = r_{k_0}$, we have $S(k_0, r)\backslash\{k_0\} \subset G\backslash\mathfrak{N}$). This implies that the formula

$$u(k) = \psi(k) - (k - k_0)^{-1}\psi_{-1}$$

enables us to extend $u(k)$ with $S(k_0, r_1)$ analytically onto the larger circle $S(k_0, r)$. Statement [ii] of Theorem 16 now follows from the form of ψ_{-1} (formula (42)) and from the equation $\alpha_0 = \alpha(k_0)$.

Now let us prove [i]. Let $g_0 = 1$, $k_0 = 0 \in \mathfrak{N} \cap G$. We have already determined the number $r = r_{k_0} = r_0$ for which $S(0, r) \subset G$ and $S(0, r) \cap \mathfrak{N} = \{0\}$. As before, let r' be a positive number such that $S(0, r') \subset G$ and such that for $k \in S(0, r')$ we have

$$\alpha(k) = \sum_{\nu=0}^{\infty} k^{\nu}\alpha_{\nu}; \quad \alpha_{\nu} \in L_2(\Omega), \quad \|\alpha_{\nu}\| \leqslant C_0 (r')^{-\nu},$$

$\nu = 0, 1, 2, \cdots$; $C_0 = \text{const} > 0$. We set

$$A_0 = T(0) + \frac{3}{2\sqrt{7\pi}} \cdot \frac{1}{1 - g_2},$$

and choose the number r_1 in such a way that $0 < r_1 < \min\{r, r', A_0^{-1}\}$.

Let $\alpha_{-1} = \vartheta$. Let us incorporate the sequence $\{\psi_{\nu} | \nu \geq -2\} \subset L_2(\Omega)$ into the sequence $\{\alpha_{\nu} | \nu = -1, 0, 1, 2, \cdots\}$ in accordance with Lemma 18. Then (compare (45))

$$\|\psi_{-2}\| \leqslant \frac{3}{2\sqrt{\pi}}(1 - g_1)\|\alpha_0\|.$$

It follows from (47) and (48) that for $\nu \geq -1$ we have

$$\|\psi_{\nu}\| \leqslant T(0)(\|\alpha_{\nu}\| + \|\psi_{\nu-1}\|) + \frac{3}{2\sqrt{\pi}}(1 - g_1)\|\alpha_{\nu+2}\|$$

$$+ \frac{1}{2}\sqrt{\frac{3}{\pi}}\|\alpha_{\nu+1}\| + \frac{3}{2\sqrt{5\pi}}\frac{1}{1 - g_2}\|\alpha_{\nu}\| + \frac{3}{2\sqrt{7\pi}} \cdot \frac{1}{1 - g_2}\|\psi_{\nu-1}\| \leqslant A_0\|\psi_{\nu-1}\| + \frac{B^0}{r_1^{\nu}},$$

where

$$B^0 = C_0\left\{T(0) + \frac{3}{2\sqrt{5\pi}}\frac{1}{1 - g_2} + \frac{1}{2r_1}\sqrt{\frac{3}{\pi}} + \frac{3}{2\sqrt{\pi}}\frac{1 - g_1}{r_1^2}\right\}.$$

Exactly as we did in proving [ii], we can use the above inequalities to obtain the estimate

$$\|\psi_\nu\| \leqslant \frac{D_0}{r_1^\nu}, \quad \nu \geqslant -1;$$

$$D_0 = 2\max\left\{\frac{3}{2\sqrt{\pi}}(1-g_1)A_0^2C_0, \quad \frac{B^0}{1-A_0r_1}\right\}.$$

The estimate ensures convergence in $L_2(\Omega)$ of the power series

$$\psi^{(0)}(\mathrm{k}) = \sum_{\nu=-2}^{\infty} \mathrm{k}^\nu \psi_\nu, \quad 0 < |\mathrm{k}| < r_1.$$

Let $u(\mathrm{k}) = \sum_{\nu=0}^{\infty} \mathrm{k}^\nu \psi_\nu$. The function $u(\mathrm{k})$ is holomorphic in $S(0, r_1)$, and for $\mathrm{k} \in S(0, r_1)$, $\mathrm{k} \neq 0$, we have

$$\psi^{(0)}(\mathrm{k}) = \mathrm{k}^{-2}\psi_{-2} + \mathrm{k}^{-1}\psi_{-1} + u(\mathrm{k}).$$

The coefficients ψ_ν satisfy system (44). This implies that $\psi^{(0)}(\mathrm{k})$ is a solution of equation (39), $\mathrm{k} \in S(0, r_1)$, $\mathrm{k} \neq 0$:

$$(1 + \mathrm{k}\boldsymbol{\omega}\mathbf{n})\,\psi^{(0)}(\mathrm{k}) = \psi^{(0)}(\mathrm{k}) + (\boldsymbol{\omega}\mathbf{n})\,\mathrm{k}\psi^{(0)}(\mathrm{k})$$

$$= \sum_{\nu=-2}^{\infty} \mathrm{k}^\nu\psi_\nu + (\boldsymbol{\omega}\mathbf{n}) \sum_{\nu=-1}^{\infty} \mathrm{k}^\nu\psi_{\nu-1} = \mathrm{k}^{-2}\psi_{-2} + \sum_{\nu=-1}^{\infty} \mathrm{k}^\nu(\psi_\nu + (\boldsymbol{\omega}\mathbf{n})\,\psi_{\nu-1})$$

$$= \mathrm{k}^{-2}\hat{g}\psi_{-2} + \sum_{\nu=-1}^{\infty} \mathrm{k}^\nu(\hat{g}\psi_\nu + \alpha_\nu) = \hat{g}\psi^{(0)}(\mathrm{k}) + \alpha(\mathrm{k}).$$

Because (39) is uniquely solvable, for $\mathrm{k} \in S(0, r_1)\backslash\{0\} \subset S(0, r)\backslash\{0\} \subset G\backslash\mathfrak{N}$ we have

$$u(\mathrm{k}) = \psi^{(0)}(\mathrm{k}) - \mathrm{k}^{-2}\psi_{-2} - \mathrm{k}^{-1}\psi_{-1} = \psi(\mathrm{k}) - \mathrm{k}^{-2}\psi_{-2} - \mathrm{k}^{-1}\psi_{-1}.$$

But $\psi(\mathrm{k})$ is holomorphic in $S(0, r)\backslash\{0\}$. Hence $u(\mathrm{k})$ is holomorphic in $S(0, r)$. All that remains to prove [i] is to make use of formulas (45) and (51) and to note that $\alpha_0 = \alpha(0)$, $\alpha_1 = \alpha'(0) \equiv (d/d\mathrm{k})\,\alpha(\mathrm{k})|_{\mathrm{k}=0}$.

§7. The operator $\hat{A}$

In this section we turn to the investigation of the actual problem of the passage of radiation through a semi-infinite plane material layer. In our Introduction (§1) we reduced this problem to the equation

$$x(\tau, \boldsymbol{\omega}) = \hat{A}x(\tau, \boldsymbol{\omega}) + f(\tau, \boldsymbol{\omega}), \quad \tau \in (0, \infty), \quad \boldsymbol{\omega} \in \Omega, \tag{52}$$

with the special operator $\hat{A}$ defined by (4). We must now isolate the classes of functions $x(\tau, \omega)$ on which the operator $\hat{A}$ can be defined most naturally and investigate its properties. We can do this most conveniently with the aid of the notions of measurability and summability in the sense of Bochner [35] of abstract functions of a real variable.

Let X be some separable Banach space of numerical functions $f(\omega)$ defined on the sphere Ω (e.g. $X = L_2(\Omega)$, $X = C(\Omega)$); let X^* be the corresponding adjoint space. We shall consider the abstract functions $x(\tau)$ of the real variable $\tau \in (0, \infty)$ with values in X. In order to distinguish such a function from its value $x(\tau) \in X$ at the point τ, we shall usually denote it by $x(\cdot)$. By $x(\tau, \omega)$ we denote the numerical value assumed by the function $x(\tau)$ at the point $\omega \in \Omega$.

We require that the functions $x(\cdot)$ be strongly Bochner-measurable [35] with respect to the independent variable τ. Because X is separable, the function $x(\cdot)$ is strongly measurable if and only if all numerical functions of the form $x^*(x(\tau))$, where x^* is an arbitrary fixed element of the space X^*, are Lebesgue-measurable [35]. It turns out that the numerical function $\|x(\tau)\|$ is Lebesgue-measurable for every strongly measurable function $x(\cdot)$; a linear combination of strongly measurable functions with constant coefficients is strongly measurable; the product of a strongly measurable function and a numerical measurable function is strongly measurable, etc.

The strongly measurable functions $x(\cdot)$ for which the integral $\int_0^\infty \|x(\tau)\| d\tau$ is finite form a class of functions Bochner-summable (B-summable) on $(0, \infty)$. To each such function $x(\cdot)$ we can in some unambiguous way [35] assign an element of the space X which we denote by $\int_0^\infty x(\tau)d\tau$. The Bochner integral $\int_0^\infty x(\tau)d\tau$ is additive and homogeneous with respect to multiplication of the function $x(\cdot)$ by complex numbers. If x^* is a fixed element of X^*, and if $x(\cdot)$ is a function B-summable on $(0, \infty)$ with values from X, then

$$x^*\left(\int_0^\infty x(\tau)\,d\tau\right) = \int_0^\infty x^*(x(\tau))\,d\tau, \tag{53}$$

$$\left\|\int_0^\infty x(\tau)\,d\tau\right\| \leqslant \int_0^\infty \|x(\tau)\|\,d\tau. \tag{54}$$

Throughout the discussion to follow, as in the previous sections, n is a fixed unit vector. Its geometric meaning was explained in the Introduction: n is the interior normal to the boundary of the halfspace $\tau > 0$. The function g in this section will be subject to the single restriction $g \in L_2(-1, 1)$; the operator $\hat{g}: L_2(\Omega) \to L_2(\Omega)$ will have the same meaning as in Theorem 1.

Lemma 19. *Let $g \in L_2(-1, 1)$, and let the operator $\hat{g}: L_2(\Omega) \to L_2(\Omega)$ be defined as in Theorem 1. Let $x(\cdot)$ be an abstract function of the variable $\tau \in (0, \infty)$ with values in $L_2(\Omega)$, and let $y(\cdot)$ be a function defined by the relation*

$$y(\tau) = \hat{g}x(\tau), \quad \tau \in (0, \infty).$$

Then $y(\tau) \in C(\Omega)$ and $\|y(\tau)\|_{C(\Omega)} \leq \sqrt{2\pi}\,[g]\|x(\tau)\|$. If $x(\cdot)$ is strongly measurable (B-summable), then $y(\cdot)$ considered as a function of $\tau \in (0, \infty)$ with values in *space* $C(\Omega)$ *is also strongly measurable* (B-*summable*).

Proof. To each point $\omega \in \Omega$ we assign a functional x^*_ω acting in $C(\Omega)$ according to the formula

$$x^*_\omega(x) = x(\omega) \quad \text{for} \quad x \in C(\Omega). \tag{55}$$

Clearly, $x^*_\omega \in C^*(\Omega)$. Because Ω is compact, there exists a countable set $\Lambda \subset \Omega$ such that the linear shell of the corresponding system of functionals $\Lambda^* = \{x^*_\omega | \omega \in \Lambda\}$ is weakly dense in $C^*(\Omega)$.

Let $x(\cdot)$ be a strongly measurable function of the variable $\tau \in (0, \infty)$ with values in $L_2(\Omega)$. The inclusion $y(\tau) = \hat{g}x(\tau) \in C(\Omega)$ and the inequality

$$\| y(\tau) \|_{C(\Omega)} \leqslant \sqrt{2\pi}\, [g] \, \| x(\tau) \|, \quad \tau \in (0, \infty),$$

are ensured by Theorem 1. We choose an $\omega_0 \in \Omega$ and let $u(\omega) = \overline{g(\omega\omega_0)}$. Then $u \in L_2(\Omega)$ (Lemma 1) and, because of the strong measurability of $x(\cdot)$, the numerical function $(x(\tau), u)$ of the variable τ is Lebesgue-measurable on $(0, \infty)$. But

$$(x(\tau), u) = y(\tau, \omega_0) = x^*_{\omega_0}(y(\tau)).$$

Specifically, $x^*(y(\tau))$ is a measurable numerical function of τ for every $x^* \in \Lambda^*$.

Now let x^* be an arbitrary element of $C^*(\Omega)$ and let $\{x^*_s | s = 1, 2, \cdots\}$ be a sequence of linear combinations of elements of Λ^* which converges to this element. According to what we said above, $x^*_s(y(\tau))$ is a measurable function of $\tau \in (0, \infty)$, $s = 1, 2, \cdots$. On the other hand, $x^*(y(\tau)) = \lim_{s=\infty} x^*_s(y(\tau))$ for every $\tau \in (0, \infty)$. This means that $x^*(y(\tau))$ is measurable on $(0, \infty)$. The strong measurability of the abstract function $y(\cdot)$ with values in $C(\Omega)$ now follows from the arbitrariness of $x^* \in C^*(\Omega)$.

If $x(\cdot)B$ is summable on $(0, \infty)$, then the above estimate for $\|y(\tau)\|_{C(\Omega)}$ implies the finiteness of the integral $\int_0^\infty \|y(\tau)\|_{C(\Omega)} d\tau$ and also the B-summability of $y(\cdot)$ (as a function with values in $C(\Omega)$). The lemma is proved.

Now let us isolate the classes of functions $x(\tau, \omega)$ in which we shall be conducting our discussion.

Definition. Let $\mathcal{K}$ be the class of all functions $x(\cdot)$ of the variable $\tau \in (0, \infty)$ with values in $L_2(\Omega)$, and let ϵ be a real number. Next, we denote by $\mathcal{B}_\epsilon$ the set of abstract functions $x(\cdot)$ having the following properties:

a) $x(\cdot) \in \mathcal{K}$.

b) The numerical function $\|x(\tau)\|$ is bounded on every bounded portion of the half-line $(0, \infty)$.

c) The function $e^{-\epsilon\tau}x(\tau)$ is B-summable on $(0, \infty)$.

It is clear that $\mathfrak{B}_\epsilon \subset \mathfrak{B}_{\epsilon'} \subset \mathfrak{K}$ for $\epsilon > \epsilon'$. We set

$$\overline{\mathfrak{B}} = \bigcup_{\varepsilon<1} \mathfrak{B}_\varepsilon. \tag{56}$$

Then $\mathfrak{B}_\epsilon \subset \overline{\mathfrak{B}} \subset \mathfrak{B}_1$, $\epsilon < 1$.

It is clear that the class $\mathfrak{K}$ is a linear system of relatively naturally defined operators of multiplication by complex numbers and addition of the elements of $\mathfrak{K}$. Each set $\mathfrak{B}_\epsilon$ is a linear manifold in $\mathfrak{K}$.

The class $\mathfrak{B}_1$ turns out to be the natural domain of definition of the operator $\hat{A}$ in accordance with formulas (4). In order to show this we shall carry out some ancillary constructions.

Let $x(\cdot) \in \mathfrak{B}_1$ and $y(\tau) = \hat{g}x(\tau)$, $\tau \in (0, \infty)$. By Lemma 19, $y(\cdot)$ is an abstract strongly measurable function of the argument $\tau \in (0, \infty)$ with values in $C(\Omega)$; moreover $\sup_{\tau \in (0,\tau_1)} \|y(\tau)\|_{C(\Omega)} < \infty$ for every $\tau_1 > 0$; finally the function $e^{-\tau}y(\tau)$ is B-summable on $(0, \infty)$. This implies that for every $\omega \in \Omega$ the numerical function $x^*(y(\tau))$ is measurable and (since $|x^*_\omega(y(\tau))| \le \|y(\tau)\|_{C(\Omega)}$) summable on $(0, \infty)$ with the weight $e^{-\tau}$. This enables us to define the numerical functions $z_\pm(\tau, \omega)$ of the pair of variables $(\tau, \omega) \in (0, \infty) \times \Omega$:

$$z_+(\tau, \omega) = \begin{cases} \dfrac{1}{\omega n} \displaystyle\int_0^\tau e^{-\frac{\tau-\rho}{\omega n}} \overset{*}{x}_\omega (\hat{g}x(\rho))\, d\rho & \text{for } \omega n > 0; \\ 0 & \text{for } \omega n \leqslant 0. \end{cases} \tag{57}$$

$$z_-(\tau, \omega) = \begin{cases} \dfrac{0}{|\omega n|} \displaystyle\int_\tau^\infty e^{-\frac{\rho-\tau}{|\omega n|}} \overset{*}{x}_\omega (\hat{g}x(\rho))\, d\rho & \text{for } \omega n < 0; \\ 0 & \text{for } \omega n \geqslant 0. \end{cases} \tag{58}$$

The numbers $z_\pm(\tau, \omega)$ are defined uniquely by specifying the function $x(\cdot) \in \mathfrak{B}_1$ and the point (τ, ω) and do not vary with changes in the values of $x(\rho)$ on the set of points ρ of zero measure.

Along with $z_\pm(\tau, \omega)$ we introduce the function

$$z_0(\tau, \omega) = \begin{cases} \overset{*}{x}_\omega (\hat{g}x(\tau)) & \text{for } \omega n = 0, \\ 0 & \text{for } \omega n \neq 0. \end{cases} \tag{59}$$

If $x(\cdot) \in \mathfrak{B}_1$, then, by the definition of $\mathfrak{B}_1$, $x(\tau) \in L_2(\Omega)$ for all $\tau \in (0, \infty)$. Hence the quantity $z_0(\tau, \omega)$ is defined uniquely by formula (59) for all $(\tau, \omega) \in (0, \infty) \times \Omega$.

The functions $z_{\pm,0}$ can also be regarded as abstract functions. In accordance with the above notation, $z_{\pm,0}(\tau)$ is some function defined on the sphere Ω; it depends on τ as a parameter and assumes the numerical value $z_{\pm,0}(\tau, \omega)$ at the point $\omega \in \Omega$. The function $z_{\pm,0}(\cdot)$ is, in turn, an abstract function of the variable $\tau \in (0, \infty)$ whose value at the fixed point τ is a function on the sphere $z_{\pm,0}(\tau)$.

Thus, on the basis of the given function $g \in L_2(-1, 1)$ we have constructed the single-valued correspondence

$$x(\cdot) \longrightarrow \{z_+(\cdot), z_-(\cdot), z_0(\cdot)\}, \; x(\cdot) \in \mathfrak{B}_1, \tag{60}$$

which converts each function $x(\cdot) \in \mathfrak{B}_1$ into a triplet of new abstract functions. The following lemma summarizes the properties of these functions.

Lemma 20. *Let $x(\cdot) \in \mathfrak{B}_1$, and let the functions $z_{\pm,0}(\cdot)$ be defined in accordance with* (60). *Then*

[i] *For every $\tau > 0$ we have*

$$z_+(\tau) \in L_2(\Omega), \quad z_-(\tau) \in L_2(\Omega), \quad z_0(\tau) = \vartheta \in L_2(\Omega).$$

[ii] *If $\tau_1 > 0$, then there exists a number $l(\tau_1)$, $0 \le l(\tau_1) < \infty$, such that*

$$|z_{\pm,0}(\tau, \omega)| \leqslant l(\tau_1), \quad (\tau, \omega) \in (0, \tau_1] \times \Omega;$$

$$\|z_\pm(\tau)\| \leqslant \sqrt{2\pi} l(\tau_1), \quad \tau \in (0, \tau_1].$$

[iii] *For a fixed $\tau \in (0, \infty)$ the functions $z_{\pm,0}(\tau, \omega)$ are continuous with respect to ω at every point $\omega_0 \in \Omega$ for which $\omega_0 \mathbf{n} \neq 0$.*

[iv] *If $\delta > 0$, $\tau > 0$, then*

$$\lim_{\tau' = \tau} z_+(\tau', \omega) = z_+(\tau, \omega),$$

uniformly with respect to all $\omega \in \Omega$ for which $\omega \mathbf{n} \,\overline{\in}\, (0, \delta)$, and

$$\lim_{\tau' = \tau} z_-(\tau', \omega) = z_-(\tau, \omega)$$

uniformly with respect to all $\omega \in \Omega$ for which $\omega \mathbf{n} \,\overline{\in}\, (-\delta, 0)$.

[v] *$z_{\pm,0}(\cdot)$ are continuous abstract functions of τ: for $\tau > 0$ we have*

$$\lim_{\tau' = \tau} \| z_{\pm,0}(\tau') - z_{\pm,0}(\tau) \| = 0.$$

[vi] *For $\tau > 0$,*

$$\|z_0(\tau)\| = 0;$$

$$\|z_+(\tau)\| \leqslant \pi \sqrt{2}\, [g] \int_0^\tau e^{-(\tau-\rho)} \frac{\|x(\rho)\|}{\sqrt{\tau - \rho}}\, d\rho;$$

$$\|z_-(\tau)\| \leqslant \pi \sqrt{2}\, [g] \int_\tau^\infty e^{-(\rho-\tau)} \frac{\|x(\rho)\|}{\sqrt{\rho - \tau}}\, d\rho.$$

Proof. First let us prove [ii]. Let $l_0(\tau) = \sup_{\tau' \in (0,\tau]} \|x(\tau')\|$, $\tau > 0$. By the definition of $\mathcal{B}_1$, $0 \le l_0(\tau) < \infty$, $l_0(\tau)$ does not decrease with increasing τ. If $\tau_1 > 0$, then for $\tau \in (0, \tau_1]$, $\omega \in \Omega$, we have

$$|z_0(\tau, \omega)| \leqslant \|\hat{g}x(\tau)\|_{C(\Omega)} \leqslant \sqrt{2\pi}\,[g]\,\|x(\tau)\| \leqslant \sqrt{2\pi}\,[g]\,l_0(\tau_1).$$

For $\tau \in (0, \tau_1]$, $\omega \in \Omega$, $\omega\mathbf{n} > 0$ we have

$$|z_+(\tau, \omega)| \leqslant \frac{1}{\omega\mathbf{n}} \int_0^\tau e^{-\frac{\tau-\rho}{\omega\mathbf{n}}} \|\hat{g}x(\rho)\|_{C(\Omega)}\,d\rho$$

$$\leqslant \sqrt{2\pi}\,[g]\,l_0(\tau_1) \frac{1}{\omega\mathbf{n}} \int_0^\tau e^{-\frac{\tau-\rho}{\omega\mathbf{n}}}\,d\rho \leqslant \sqrt{2\pi}\,[g]\,l_0(\tau_1).$$

But if $\tau \in (0, \tau_1]$, $\omega \in \Omega$, $\omega\mathbf{n} < 0$, then

$$|z_-(\tau, \omega)| \leqslant \frac{\sqrt{2\pi}\,[g]}{|\omega\mathbf{n}|} \int_\tau^\infty e^{-\frac{\rho-\tau}{|\omega\mathbf{n}|}} \|x(\rho)\|\,d\rho$$

$$= \frac{\sqrt{2\pi}\,[g]}{|\omega\mathbf{n}|} \int_\tau^{\tau+1} e^{-\frac{\rho-\tau}{|\omega\mathbf{n}|}} \|x(\rho)\|\,d\rho + \frac{\sqrt{2\pi}\,[g]}{|\omega\mathbf{n}|} \int_{\tau+1}^\infty e^{-\frac{\rho-\tau}{|\omega\mathbf{n}|}} \|x(\rho)\|\,d\rho$$

$$\leqslant \frac{\sqrt{2\pi}\,[g]}{|\omega\mathbf{n}|}\, l_0(\tau_1+1) \int_\tau^{\tau+1} e^{-\frac{\rho-\tau}{|\omega\mathbf{n}|}}\,d\rho + \frac{\sqrt{2\pi}\,[g]}{|\omega\mathbf{n}|}\, e^{-\frac{1}{|\omega\mathbf{n}|}} \int_{\tau+1}^\infty e^{-(\rho-(\tau+1))} \|x(\rho)\|\,d\rho$$

$$\leqslant \sqrt{2\pi}\,[g]\,l_0(\tau_1+1)\left(1 - e^{-\frac{1}{|\omega\mathbf{n}|}}\right) + \sqrt{2\pi}\,[g]\,e^{\tau} \int_{\tau+1}^\infty e^{-\rho} \|x(\rho)\|\,d\rho$$

$$\leqslant \sqrt{2\pi}\,[g] \left\{ l_0(\tau_1+1) + e^{\tau_1} \int_1^\infty e^{-\rho} \|x(\rho)\|\,d\rho \right\} \equiv l(\tau_1) < \infty.$$

According to the above estimates, for $(\tau, \omega) \in (0, \tau_1] \times \Omega$ we have

$$|z_{\pm,0}(\tau, \omega)| \leqslant l(\tau_1).$$

Every function $z_\pm(\tau, \omega)$ of the variable $\omega \in \Omega$ vanishes on a hemisphere. Hence

$$\|z_\pm(\tau)\|^2 = \int_\Omega |z_\pm(\tau, \omega)|^2\,d\omega \leqslant l^2(\tau_1) \int_{\omega\mathbf{n} \gtrless 0} d\omega = 2\pi l^2(\tau_1), \quad \tau \in (0, \tau_1].$$

(The measurability of $z_\pm(\tau, \omega)$ with respect to ω on Ω will be established below.) [ii] has been proved.

Now let us turn to [iii]. Let $\tau > 0$, $\omega_0 \in \Omega$, $\omega_0\mathbf{n} < 0$. To prove the continuity of $z_-(\tau, \omega)$ with respect to ω at the point ω_0 we need only justify the limit relation $\lim_{\omega=\omega_0} d = 0$, where

$$d = d(\omega) = \left| \int_\tau^\infty x_\omega^*(y(\rho))\, e^{-\frac{\rho}{|\omega n|}}\, d\rho - \int_\tau^\infty x_{\omega_0}^*(y(\rho))\, e^{-\frac{\rho}{|\omega_0 n|}}\, d\rho \right|.$$

$$y(\rho) = \hat{g}x(\rho).$$

We have

$$d \leqslant \int_\tau^\infty e^{-\rho} \left| e^{-\rho(|\omega n|^{-1}-1)} - e^{-\rho(|\omega_0 n|^{-1}-1)} \right| |x_\omega^*(y(\rho))|\, d\rho$$

$$+ \int_\tau^\infty e^{-\rho/|\omega_0 n|} |x_\omega^*(y(\rho)) - x_{\omega_0}^*(y(\rho))|\, d\rho$$

$$\leqslant \sqrt{2\pi}\,[g] \int_\tau^\infty e^{-\rho} \|x(\rho)\| \left| e^{-\rho(|\omega n|^{-1}-1)} - e^{-\rho(|\omega_0 n|^{-1}-1)} \right| d\rho$$

$$+ \int_\tau^\infty e^{-\rho} |x_\omega^*(y(\rho)) - x_{\omega_0}^*(y(\rho))|\, d\rho.$$

The integrand of the first integral is dominated by the summable function $2e^{-\rho}\|x(\rho)\|$. That of the second integral is dominated by the summable function $2\sqrt{2\pi}\,[g]e^{-\rho}\|x(\rho)\|$. For a fixed $\rho > \tau$ both integrands tend to zero as ω tends to ω_0. This is self-evident in the former case; in the latter it follows from the fact that $x_\omega^*(y(\rho)) = y(\rho, \omega)$ and from the inclusion $y(\rho) \in C(\Omega)$. Hence as $\omega \to \omega_0$ each of the integrals tends to zero (by the Lebesgue theorem on dominated convergence), $\lim_{\omega=\omega_0} d = 0$ and $z_-(\tau, \omega)$ is continuous at the point ω_0. The continuity of $z_{+,0}(\tau, \omega)$ at the point ω_0, $\omega_0 n < 0$, follows from the definition of these functions.

Let $\omega_0 n > 0$. The continuity of the functions $z_{-,0}(\tau, \omega)$ at the point ω_0 is self-evident, and to prove the continuity of $z_+(\tau, \omega)$ we need merely prove the limit relation

$$\lim_{\omega=\omega_0} \int_0^\tau e^{-\frac{\tau-\rho}{\omega n}} x_\omega^*(y(\rho))\, d\rho = \int_0^\tau e^{-\frac{\tau-\rho}{\omega_0 n}} x_{\omega_0}^*(y(\rho))\, d\rho.$$

For all ω sufficiently close to ω_0 the integrand of the left-hand integral is bounded by the constant $\sqrt{2\pi}\,[g]l_0(\tau) < \infty$; for every fixed $\rho \in (0, \tau)$ we have

$$\lim_{\omega=\omega_0} e^{-\frac{\tau-\rho}{\omega n}} x_\omega^*(y(\rho)) = e^{-\frac{\tau-\rho}{\omega_0 n}} x_{\omega_0}^*(y(\rho)).$$

Again applying the Lebesgue theorem, we arrive at [iii].

Statement [iii] implies the measurability of the functions $z_{\pm,0}(\tau, \omega)$ with respect to ω on Ω. This and [ii] imply statement [i].

We choose $\tau_1 > 0$, $\omega \in \Omega$, $\omega n > 0$. Let $0 < \tau < \tau' < \tau_1$.

$$|z_+(\tau,\omega)-z_+(\tau',\omega)|\leqslant\frac{\sqrt{2\pi}\,[g]}{\omega n}\int_\tau^{\tau'}\|x(\rho)\|e^{-\frac{\tau'-\rho}{\omega n}}\,d\rho$$

$$+\frac{\sqrt{2\pi}\,[g]}{\omega n}\int_0^\tau\|x(\rho)\|\left(e^{-\frac{\tau-\rho}{\omega n}}-e^{-\frac{\tau'-\rho}{\omega n}}\right)d\rho\leqslant\sqrt{2\pi}\,[g]\,l_0(\tau_1)\left(1-e^{-\frac{\tau'-\tau}{\omega n}}\right)$$

$$+\sqrt{2\pi}\,[g]\,l_0(\tau_1)\left(1-e^{-\frac{\tau}{\omega n}}-e^{-\frac{\tau'-\tau}{\omega n}}+e^{-\frac{\tau'}{\omega n}}\right)$$

$$\leqslant 2\sqrt{2\pi}\,[g]\,l_0(\tau_1)\left(1-e^{-\frac{\tau'-\tau}{\omega n}}\right).$$

If $\delta>0$, $0<\tau<\tau'<\tau_1$, and if $\omega\in\Omega$ is such that $\omega n \,\overline{\in}\,(0,\delta)$, then

$$|z_+(\tau,\omega)-z_+(\tau',\omega)|\leqslant 2\sqrt{2\pi}\,[g]\,l_0(\tau_1)\left(1-e^{-\frac{\tau'-\tau}{\delta}}\right),$$

so that $\lim_{\tau'=\tau} z_+(\tau',\omega)=z_+(\tau',\omega)$ uniformly with respect to all $\omega\in\Omega$ satisfying the condition $\omega n\,\overline{\in}\,(0,\delta)$.

Now let us assume that $\tau_1>0$, $\omega\in\Omega$, $\omega n>0$. For $0<\tau<\tau'<\tau_1$ we have

$$|z_-(\tau,\omega)-z_-(\tau',\omega)|\leqslant\frac{\sqrt{2\pi}\,[g]}{|\omega n|}\int_\tau^{\tau'}e^{-\frac{\rho-\tau}{|\omega n|}}\|x(\rho)\|\,d\rho$$

$$+\frac{\sqrt{2\pi}\,[g]}{|\omega n|}\int_{\tau'}^{\tau_1+2}\|x(\rho)\|\left(e^{-\frac{\rho-\tau'}{|\omega n|}}-e^{-\frac{\rho-\tau}{|\omega n|}}\right)d\rho$$

$$+\frac{\sqrt{2\pi}\,[g]}{|\omega n|}\int_{\tau_1+2}^{\infty}\|x(\rho)\|\left(e^{-\frac{\rho-\tau'}{|\omega n|}}-e^{-\frac{\rho-\tau}{|\omega n|}}\right)d\rho\leqslant\sqrt{2\pi}\,[g]\,l_0(\tau_1)\left(1-e^{-\frac{\tau'-\tau}{|\omega n|}}\right)$$

$$+\sqrt{2\pi}\,[g]\,l_0(\tau_1+2)\left[1-e^{-\frac{\tau_1+2-\tau'}{|\omega n|}}-e^{-\frac{\tau'-\tau}{|\omega n|}}+e^{-\frac{\tau_1+2-\tau}{|\omega n|}}\right]$$

$$+\sqrt{2\pi}\,[g]\frac{1}{|\omega n|}e^{-\frac{1}{|\omega n|}}\int_{\tau_1+2}^{\infty}\|x(\rho)\|\left(e^{-\frac{\rho-\tau'-1}{|\omega n|}}-e^{-\frac{\rho-\tau-1}{|\omega n|}}\right)d\rho$$

$$\leqslant 2\sqrt{2\pi}\,[g]\,l_0(\tau_1+2)\left(1-e^{-\frac{\tau'-\tau}{|\omega n|}}\right)+$$

$$+\frac{\sqrt{2\pi}\,[g]}{e}\int_{\tau_1+2}^{\infty}\|x(\rho)\|\left(e^{-\frac{\rho-\tau'-1}{|\omega n|}}-e^{-\frac{\rho-\tau-1}{|\omega n|}}\right)d\rho.$$

Let us choose a $\rho\geq\tau_1+2$, and let

$$a=\rho-\tau-1,\quad a'=\rho-\tau'-1,\quad \xi=1/|\omega n|.$$

Clearly, $a>a'>1$, $\xi\geq 1$. Therefore

$$e^{-\frac{\rho-\tau'-1}{|\omega n|}}-e^{-\frac{\rho-\tau-1}{|\omega n|}}=e^{-a'\xi}-e^{-a\xi}\leqslant e^{-a'}-e^{-a}=e^{-(\rho-\tau'-1)}-e^{-(\rho-\tau-1)}.$$

Hence for $\omega n < 0$, $0 < \tau < \tau' < \tau_1$ we have

$$|z_-(\tau', \omega) - z_-(\tau, \omega)| \leqslant 2\sqrt{2\pi}\,[g]\, l_0(\tau_1 + 2)\left(1 - e^{-\frac{\tau'-\tau}{|\omega n|}}\right)$$
$$+ \sqrt{2\pi}\,[g]\,(e^{\tau'} - e^{\tau}) \int_{\tau_1+2}^{\infty} e^{-\rho}\|x(\rho)\|\,d\rho.$$

If $\delta > 0$, $0 < \tau < \tau' < \tau_1$, and if $\omega \in \Omega$ is such that $\omega n \,\bar{\in}\, (-\delta, 0)$, then

$$|z_-(\tau', \omega) - z_-(\tau, \omega)| \leqslant 2\sqrt{2\pi}\,[g]\, l_0(\tau_1 + 2)\left(1 - e^{-\frac{\tau'-\tau}{\delta}}\right).$$
$$+ \sqrt{2\pi}\,[g]\,(e^{\tau'} - e^{\tau}) \int_{\tau_1+2}^{\infty} e^{-\rho}\|x(\rho)\|\,d\rho,$$

and $\lim_{\tau'=\tau} z_-(\tau', \omega) = z_-(\tau, \omega)$ uniformly with respect to all $\omega \in \Omega$ satisfying the conditions $\omega n \,\bar{\in}\, (-\delta, 0)$. [iv] has been proved.

To prove [v] we choose $\tau_1 > 0$ and let $0 < \tau < \tau_1$, $0 < \tau' < \tau_1$. Then

$$\|z_\pm(\tau) - z_\pm(\tau')\|^2 = \int_\Omega |z_\pm(\tau, \omega) - z_\pm(\tau', \omega)|^2\, d\omega.$$

The integrand is bounded (compare [ii]):

$$|z_\pm(\tau, \omega) - z_\pm(\tau', \omega)| \leqslant 2l(\tau_1), \quad \tau \in (0, \tau_1), \quad \tau' \in (0, \tau_1), \quad \omega \in \Omega.$$

At the same time, according to [iv], for all $\omega \in \Omega$, we have

$$\lim_{\tau'=\tau} |z_\pm(\tau', \omega) - z_\pm(\tau, \omega)| = 0.$$

This implies the limit relation

$$\lim_{\tau'=\tau} \|z_\pm(\tau') - z_\pm(\tau)\| = 0.$$

Finally, for $\tau > 0$ and $\tau' > 0$ we have

$$\|z_0(\tau') - z_0(\tau)\| = 0.$$

It remains for us to justify estimates [vi]. For $\tau > 0$, $\omega \in \Omega$, $\omega n < 0$ we have

$$|z_-(\tau, \omega)| \leqslant \frac{\sqrt{2\pi}\,[g]}{|\omega n|} \int_\tau^\infty e^{-\frac{\rho-\tau}{|\omega n|}} \|x(\rho)\|\,d\rho.$$

This means that

$$|z_-(\tau, \omega)|^2 \leqslant \frac{2\pi\,[g]^2}{|\omega n|^2} \int_\tau^\infty \int_\tau^\infty \|x(\rho')\|\,\|x(\rho)\|\, e^{-\frac{\rho'+\rho-2\tau}{|\omega n|}}\, d\rho\, d\rho'.$$

Hence (compare Lemma 1)

$$\|z_-(\tau)\|^2 = \int_\Omega |z_-(\tau,\omega)|^2 d\omega$$

$$\leqslant 4\pi^2[g]^2 \int_\tau^\infty\int_\tau^\infty \|x(\rho')\|\cdot\|x(\rho)\| d\rho\, d\rho' \int_{-1}^0 e^{-\frac{\rho'+\rho-2\tau}{|\mu|}} \frac{d\mu}{\mu^2}$$

$$= 4\pi^2[g]^2 \int_\tau^\infty\int_\tau^\infty \|x(\rho')\|\|x(\rho)\| d\rho\, d\rho' \cdot \frac{e^{-(\rho'-\tau)}\cdot e^{-(\rho-\tau)}}{(\rho'-\tau)+(\rho-\tau)}$$

$$\leqslant 4\pi^2[g]^2 \int_\tau^\infty\int_\tau^\infty \|x(\rho')\|\|x(\rho)\| d\rho\, d\rho' \cdot \frac{1}{2}\, \frac{e^{-(\rho'-\tau)}\cdot e^{-(\rho-\tau)}}{\sqrt{\rho'-\tau}\sqrt{\rho-\tau}}$$

$$= 2\pi^2[g]^2 \left(\int_\tau^\infty e^{-(\rho-\tau)} \frac{\|x(\rho)\|}{\sqrt{\rho-\tau}} d\rho\right)^2,$$

$$\|z_-(\tau)\| \leqslant \pi\sqrt{2}[g] \int_\tau^\infty e^{-(\rho-\tau)} \frac{\|x(\rho)\|}{\sqrt{\rho-\tau}} d\rho, \quad \tau > 0.$$

The estimate for $\|z_+(\tau)\|$ can be derived in a similar way. For $\tau > 0$, $\omega \in \Omega$, $\omega n > 0$ we have

$$|z_+(\tau,\omega)| \leqslant \frac{\sqrt{2\pi}[g]}{\omega n} \int_0^\tau e^{-\frac{\tau-\rho}{\omega n}} \|x(\rho)\| d\rho,$$

$$|z_+(\tau,\omega)|^2 \leqslant 2\pi[g]^2 \int_0^\tau\int_0^\tau \|x(\rho')\|\cdot\|x(\rho)\| d\rho\, d\rho' \left(\frac{1}{(\omega n)^2} e^{-\frac{2\tau-\rho-\rho'}{\omega n}}\right);$$

$$\|z_+(\tau)\|^2 \leqslant 4\pi^2[g]^2 \int_0^\tau\int_0^\tau \|x(\rho')\|\|x(\rho)\| d\rho\, d\rho' \int_0^1 e^{-\frac{2\tau-\rho-\rho'}{\mu}} \frac{d\mu}{\mu^2}.$$

The inner integral is

$$\frac{1}{2\tau-\rho'-\rho} e^{-(2\tau-\rho'-\rho)} \leqslant \frac{1}{2}\cdot\frac{e^{-(\tau-\rho)}}{\sqrt{\tau-\rho}}\cdot\frac{e^{-(\tau-\rho')}}{\sqrt{\tau-\rho'}},$$

from which we obtain

$$\|z_+(\tau)\| \leqslant \pi\sqrt{2}[g] \int_0^\tau e^{-(\tau-\rho)} \frac{\|x(\rho)\|}{\sqrt{\tau-\rho}} d\rho, \quad \tau > 0.$$

The equation $\|z_0(\tau)\| = 0$, $\tau > 0$, is self-evident. The lemma is proved.

Let us choose $x(\cdot) \in \mathcal{B}_1$. Correspondence (60) uniquely associates the function $x(\cdot)$ with the triplet of functions $\{z_+(\cdot), z_-(\cdot), z_0(\cdot)\}$. According to Lemma 20[i], all of these functions are elements of the class $\mathcal{K}$. Let us construct the new function

$$z(\cdot) = z_+(\cdot) + z_-(\cdot) + z_0(\cdot). \tag{61}$$

Specification of $x(\cdot) \in \mathfrak{B}_1$ determines the function $z(\cdot) \in \mathfrak{K}$ uniquely. We thus have a new single-valued correspondence conforming to the prescribed function $g \in L_2(-1, 1)$, namely

$$x(\cdot) \longrightarrow z(\cdot) \in \mathcal{K}, \quad x(\cdot) \in \mathcal{B}_1. \tag{62}$$

Definition. Let $g \in L_2(-1, 1)$. We denote by $\hat{A} \equiv \hat{A}_g$ the operator which maps $\mathfrak{B}_1$ into $\mathfrak{K}$ according to the following rule: for $x(\cdot) \in \mathfrak{B}_1$ we have

$$\hat{A}x(\cdot) = z(\cdot),$$

where $z(\cdot) \in \mathfrak{K}$ is determined by correspondence (62).

It is clear that this definition of $\hat{A}$ is a formally strict variant of the definition of $\hat{A}$ given in §1 (formula (4)).

The mapping $\hat{A}: \mathfrak{B}_1 \to \mathfrak{K}$ is linear with respect to the algebraic operations acting in $\mathfrak{K}$ (compare formulas (57)–(59)).

Theorem 17. *Let $x(\cdot) \in \mathfrak{B}_1$ and $z(\cdot) = \hat{A}x(\cdot)$. The function $z(\cdot) \in \mathfrak{K}$ has the following properties:*

[i] *$z(\tau)$ is a continuous abstract function of τ.*

[ii] *If $\tau_1 > 0$, then*

$$|z(\tau, \omega)| \leqslant l(\tau_1), \ (\tau, \omega) \in (0, \tau_1] \times \Omega;$$

$$\|z(\tau)\| \leqslant 2\sqrt{\pi} l(\tau_1), \ \tau \in (0, \tau_1]$$

($l(\tau_1)$ is the number defined in Lemma 20 [ii]).

[iii] *$z(\tau, \omega)$ depends continuously on ω on the set $\{\omega \mid \omega \in \Omega, \omega n \neq 0\}$ for $\tau > 0$.*

[iv] *Let $\tau > 0$ and $\delta > 0$. Then*

$$\lim_{\tau' \to \tau} z(\tau', \omega) = z(\tau, \omega)$$

uniformly with respect to all of the $\omega \in \Omega$ satisfying the condition $|\omega n| \geq \delta$.

[v] *If $\sup_{\tau>0} \|x(\tau)\| = k < \infty$, then*

$$|z(\tau, \omega)| \leqslant \sqrt{2\pi}\,[g]\,k, \ (\tau, \omega) \in (0, \infty) \times \Omega;$$

$$\|z(\tau)\| \leqslant 2\pi\sqrt{2}\,[g]\,k, \ \tau \in (0, \infty).$$

[vi] *If τ_0 is a strong continuity point of $x(\cdot)$, $\tau_0 > 0$, then $z(\tau_0) \in C(\Omega)$, and*

$$\lim_{\tau \to \tau_0} z(\tau, \omega) = z(\tau_0, \omega)$$

uniformly with respect to all of the $\omega \in \Omega$.

Proof. Statements [i], [ii], [iii], and [iv] follow from Lemma 20 [v], [ii], [iii], [iv], respectively. Inequalities [v] are readily obtainable from the estimate $|x^*_\omega(\hat{g}x(\rho))| \leq \sqrt{2\pi}\,[g]\,k$ and from formulas (57)–(59).

It remains for us to prove [vi]. Let us assume that the point $\tau_0 > 0$ is a point of strong discontinuity of $x(\cdot)$, i.e. that $\lim_{\tau=\tau_0} \|x(\tau) - x(\tau_0)\| = 0$. We can show that $z(\tau_0, \omega)$ is continuous with respect to ω on the sphere Ω. We already know (cf. [iii]) that $z(\tau_0, \omega)$ is continuous at every point $\omega_0 \in \Omega$, $\omega_0\mathbf{n} \neq 0$. Now let $\omega_0\mathbf{n} = 0$. We set $y(\tau) = \hat{g}x(\tau)$. Then $y(\tau) \in C(\Omega)$ for $\tau > 0$ and $\|y(\tau) - y(\tau_0)\|_{C(\Omega)} \to 0$ as $\tau \to \tau_0$.

Let $\epsilon > 0$. We choose δ_1, $0 < \delta_1 < \tau_0$ in such a way that $\|y(\tau) - y(\tau_0)\|_{C(\Omega)} < \epsilon/4$ for $|\tau - \tau_0| < \delta_1$. Next, we choose a $\delta_2 > 0$ such that

a) for $\omega_{1,2} \in \Omega$, $|\omega_1 - \omega_2| < \delta_2$,

$$|y(\tau_0, \omega_1) - y(\tau_0, \omega_2)| < \frac{\varepsilon}{4};$$

b) for $0 < \mu < \delta_1$,

$$e^{-\tau_0/\mu} < \frac{\varepsilon}{4}\left(1 + \|y(\tau_0)\|_{C(\Omega)}\right)^{-1},$$

$$\frac{1}{\mu} e^{-\frac{\delta_1}{\mu}} < \frac{\varepsilon}{4}\left\{1 + \int_0^{\tau_0} \left(\|y(\rho)\|_{C(\Omega)} + \|y(\tau_0)\|_{C(\Omega)}\right) d\rho\right\}^{-1},$$

$$\frac{1}{\mu} e^{-\frac{\delta_1}{\mu}} < \frac{\varepsilon}{4}\left\{1 + e^{\tau_0+\delta_1} \int_{\tau_0+\delta_1}^{\infty} e^{-\rho} \left(\|y(\rho)\|_{C(\Omega)} + \|y(\tau_0)\|_{C(\Omega)}\right) d\rho\right\}^{-1}.$$

It is clear that such a δ_2 exists.

Let us assume that $\omega \in \Omega$, $|\omega - \omega_0| < \delta_2$.

Three cases are possible: $\omega\mathbf{n} > 0$, $\omega\mathbf{n} < 0$, $\omega\mathbf{n} = 0$.

In the first case $(\omega\mathbf{n} > 0)$

$$0 < \omega\mathbf{n} = |(\omega - \omega_0)\,\mathbf{n}| \leqslant |\omega - \omega_0| < \delta_2$$

and

$$|z(\tau_0, \omega) - z(\tau_0, \omega_0)| = |z_+(\tau_0, \omega) - z_0(\tau_0, \omega_0)|$$

$$= \left|\frac{1}{\omega\mathbf{n}} \int_0^{\tau_0} e^{-\frac{\tau_0-\rho}{\omega\mathbf{n}}} x^*_\omega(\hat{g}x(\rho))\, d\rho - x^*_{\omega_0}(\hat{g}x(\tau_0))\right|.$$

For $\mu \neq 0$, $\tau > 0$ we have

$$e^{-\frac{\tau}{\mu}} + \frac{1}{\mu}\int_0^{\tau} e^{-\frac{\tau-\rho}{\mu}}\, d\rho = 1.$$

Hence

$$|z(\tau_0,\omega)-z(\tau_0,\omega_0)|\leqslant e^{-\frac{\tau_0}{\omega n}}\|y(\tau_0)\|_{C(\Omega)}$$

$$+\left|\frac{1}{\omega n}\int_0^{\tau_0} e^{-\frac{\tau_0-\rho}{\omega n}}(x^*_\omega(y(\rho))-x^*_{\omega_0}(y(\tau_0)))\,d\rho\right|<\frac{\varepsilon}{4}$$

$$+\frac{1}{\omega n}\int_0^{\tau_0-\delta_1} e^{-\frac{\tau_0-\rho}{\omega n}}|x^*_\omega(y(\rho))-x^*_{\omega_0}(y(\tau_0))|\,d\rho+\frac{1}{\omega n}\int_{\tau_0-\delta_1}^{\tau_0} e^{-\frac{\tau_0-\rho}{\omega n}}|x^*_\omega(y(\rho))$$

$$-x^*_{\omega_0}(y(\tau_0))|\,d\rho.$$

For $\rho\in(\tau_0-\delta_1,\tau_0)$ we have

$$|x^*_\omega(y(\rho))-x^*_{\omega_0}(y(\tau_0))|\leqslant|x^*_\omega(y(\rho))-x^*_\omega(y(\tau_0))|+|x^*_\omega(y(\tau_0))-x^*_{\omega_0}(y(\tau_0))|$$

$$\leqslant\|y(\rho)-y(\tau_0)\|_{C(\Omega)}+|y(\tau_0,\omega)-y(\tau_0,\omega_0)|<\frac{\varepsilon}{4}+\frac{\varepsilon}{4}=\frac{\varepsilon}{2}.$$

This means that

$$|z(\tau_0,\omega)-z(\tau_0,\omega_0)|<\frac{\varepsilon}{4}+\frac{\varepsilon}{2}\cdot\frac{1}{\omega n}\int_{\tau_0-\delta_1}^{\tau_0} e^{-\frac{\tau_0-\rho}{\omega n}}\,d\rho$$

$$+\frac{1}{\omega n}e^{-\frac{\delta_1}{\omega n}}\int_0^{\tau_0-\delta_1}(\|y(\rho)\|_{C(\Omega)}+\|y(\tau_0)\|_{C(\Omega)})\,d\rho<\frac{\varepsilon}{4}+\frac{\varepsilon}{2}\left(1-e^{-\frac{\delta_1}{\omega n}}\right)+\frac{\varepsilon}{4}<\varepsilon.$$

Thus, in the first case

$$|z(\tau_0,\omega)-z(\tau_0,\omega_0)|<\varepsilon.$$

Let us consider the second case ($\omega\mathbf{n}<0$). Here $0<|\omega\mathbf{n}|<\delta_2$, and

$$|z(\tau_0,\omega)-z(\tau_0,\omega_0)|=|z_-(\tau_0,\omega)-z_0(\tau_0,\omega_0)|$$

$$=\left|\frac{1}{|\omega n|}\int_{\tau_0}^{\infty} e^{-\frac{\rho-\tau_0}{|\omega n|}}[x^*_\omega(y(\rho))-x^*_{\omega_0}(y(\tau_0))]\,d\rho\right|\leqslant\frac{1}{|\omega n|}\int_{\tau_0}^{\tau_0+\delta_1} e^{-\frac{\rho-\tau_0}{|\omega n|}}|x^*_\omega(y(\rho))$$

$$-x^*_{\omega_0}(y(\tau_0))|\,d\rho+\frac{1}{|\omega n|}\int_{\tau_0+\delta_1}^{\infty} e^{-\frac{\rho-\tau_0}{|\omega n|}}|x^*_\omega(y(\rho))-x^*_{\omega_0}(y(\tau_0))|\,d\rho.$$

For $\rho\in(\tau_0,\tau_0+\delta_1)$ we have

$$|x^*_\omega(y(\rho))-x^*_{\omega_0}(y(\tau_0))|\leqslant\|y(\rho)-y(\tau_0)\|_{C(\Omega)}+|y(\tau_0,\omega)-y(\tau_0,\omega_0)|<\frac{\varepsilon}{2}.$$

Hence

$$|z(\tau_0,\omega)-z(\tau_0,\omega_0)|<\frac{\varepsilon}{2}\cdot\frac{1}{|\omega n|}\int_{\tau_0}^{\tau_0+\delta_1} e^{-\frac{\rho-\tau_0}{|\omega n|}}\,d\rho$$

$$+\frac{1}{|\omega n|}e^{-\frac{\delta_1}{|\omega n|}}\int_{\tau_0+\delta_1}^{\infty} e^{-\frac{\rho-(\tau_0+\delta_1)}{|\omega n|}}(\|y(\rho)\|_{C(\Omega)}+\|y(\tau_0)\|_{C(\Omega)})\,d\rho$$

$$\leqslant \frac{\varepsilon}{2}\left(1 - e^{-\frac{\delta_1}{|\omega n|}}\right) + \frac{1}{|\omega n|} e^{-\frac{\delta_1}{|\omega n|}} \cdot e^{\tau_0+\delta_1} \int\limits_{\tau_0+\delta_1}^{\infty} e^{-\rho} (\| y(\rho) \|_{C(\Omega)}$$

$$+ \| y(\tau_0) \|_{C(\Omega)}) \, d\rho < \frac{\varepsilon}{2} + \frac{\varepsilon}{4} < \varepsilon.$$

This means that in the second case

$$| z(\tau_0, \omega) - z(\tau_0, \omega_0) | < \varepsilon.$$

Finally, in the third case ($\omega \mathbf{n} = 0$)

$$| z(\tau_0, \omega) - z(\tau_0, \omega_0) | = | z_0(\tau_0, \omega) - z_0(\tau_0, \omega_0) |$$

$$= | x^*_\omega (y(\tau_0)) - x^*_{\omega_0} (y(\tau_0)) | = | y(\tau_0, \omega) - y(\tau_0, \omega_0) | < \frac{\varepsilon}{4} < \varepsilon.$$

Thus, whatever our $\omega \in \Omega$ satisfying the condition $|\omega - \omega_0| < \delta_2$, it is necessarily the case that

$$| z(\tau_0, \omega) - z(\tau_0, \omega_0) | < \varepsilon.$$

This completes the proof of the inclusion $z(\tau_0) \in C(\Omega)$.

Once again, let $\epsilon > 0$ and let $\delta_1 \in (0, \tau_0)$ be such that $\|y(\tau) - y(\tau_0)\|_{C(\Omega)} < \epsilon/4$ for $|\tau - \tau_0| < \delta_1$. We choose the number η in such a way that

$$0 < \eta \leqslant \frac{\delta_1}{3};$$

$$\eta \leqslant \frac{\varepsilon}{3} \left\{ 1 + \frac{9}{e^2 \delta_1^2} \int\limits_0^{\tau_0} \| y(\rho) \|_{C(\Omega)} \, d\rho + \frac{3}{2e\delta_1} \| y(\tau_0) \|_{C(\Omega)} \right\}^{-1};$$

$$\eta \leqslant \frac{\varepsilon}{3} \left\{ 1 + \frac{9}{e^2 \delta_1^2} e^{\tau_0+\delta_1} \int\limits_{\tau_0+\delta_1}^{\infty} e^{-\rho} \| y(\rho) \|_{C(\Omega)} \, d\rho + \frac{3}{2e\delta_1} \| y(\tau_0) \|_{C(\Omega)} \right\}^{-1}.$$

We can show that for $|\tau - \tau_0| < \eta$ we immediately have

$$| z(\tau, \omega) - z(\tau_0, \omega) | < \varepsilon,$$

for all $\omega \in \Omega$. Let $|\tau - \tau_0| < \eta$. If $\omega \mathbf{n} > 0$, then

$$| z(\tau, \omega) - z(\tau_0, \omega) | = \left| \frac{1}{\omega \mathbf{n}} \int\limits_0^{\tau} e^{-\frac{\tau-\rho}{\omega \mathbf{n}}} x^*_\omega (y(\rho)) \, d\rho \right.$$

$$\left. - \frac{1}{\omega \mathbf{n}} \int\limits_0^{\tau_0} e^{-\frac{\tau_0-\rho}{\omega \mathbf{n}}} x^*_\omega (y(\rho)) \, d\rho \right| \leqslant \Delta_1 + \Delta_2,$$

$$\Delta_1 = \frac{1}{\omega \mathbf{n}} \int\limits_0^{\tau_0-\delta_1} \| y(\rho) \|_{C(\Omega)} \left| e^{-\frac{\tau-\rho}{\omega \mathbf{n}}} - e^{-\frac{\tau_0-\rho}{\omega \mathbf{n}}} \right| d\rho;$$

$$\Delta_2 = \left| \frac{1}{\omega \mathbf{n}} \int\limits_{\tau_0-\delta_1}^{\tau} e^{-\frac{\tau-\rho}{\omega \mathbf{n}}} x^*_\omega (y(\rho)) \, d\rho - \frac{1}{\omega \mathbf{n}} \int\limits_{\tau_0-\delta_1}^{\tau_0} e^{-\frac{\tau_0-\rho}{\omega \mathbf{n}}} x^*_\omega (y(\rho)) \, d\rho \right|.$$

Let us estimate the first term.

$$\Delta_1 = \frac{1}{\omega n} e^{-\frac{\delta_1}{3\omega n}} \int_0^{\tau_0-\delta_1} \|y(\rho)\|_{C(\Omega)} \left| e^{-\frac{\tau - \frac{\delta_1}{3} - \rho}{\omega n}} - e^{-\frac{\tau_0 - \frac{\delta_1}{3} - \rho}{\omega n}} \right| d\rho.$$

Let $\rho \in (0, \tau_0 - \delta_1)$, $k = \tau - \delta_1/3 - \rho$, $k_0 = \tau_0 - \delta_1/3 - \rho$, $\xi = 1/\omega n$. We have

$$| e^{-k\xi} - e^{-k_0\xi} | = | k - k_0 | \xi e^{-k^*\xi},$$

where k^* is a number lying between k and k_0. But

$$k > \tau - \frac{\delta_1}{3} - \tau_0 + \delta_1 > \frac{\delta_1}{3}, \quad k_0 > \tau_0 - \frac{\delta_1}{3} - \tau_0 + \delta_1 = \frac{2\delta_1}{3},$$

in such a way that $k^* > \delta_1/3$, and

$$| e^{-k\xi} - e^{-k_0\xi} | < \frac{3}{\delta_1} | k - k_0 | (k^*\xi) e^{-k^*\xi} \leqslant \frac{3}{e\delta_1} | \tau - \tau_0 | < \frac{3\eta}{e\delta_1}.$$

Thus

$$\Delta_1 \leqslant \frac{3\eta}{e\delta_1} \cdot \frac{3}{\delta_1} \left[\frac{\delta_1}{3\omega n} e^{-\frac{\delta_1}{3\omega n}} \right] \int_0^{\tau_0-\delta_1} \|y(\rho)\|_{C(\Omega)} d\rho \leqslant \frac{9\eta}{e^2\delta_1^2} \int_0^{\tau_0} \|y(\rho)\|_{C(\Omega)} d\rho.$$

In order to estimate the second term we note that

$$\begin{aligned} \frac{1}{\omega n} \int_{\tau_0-\delta_1}^{\tau} e^{-\frac{\tau-\rho}{\omega n}} x_\omega^* (y(\rho)) \, d\rho &= x_\omega^* (y(\tau_0)) \cdot \frac{1}{\omega n} \int_{\tau_0-\delta_1}^{\tau} e^{-\frac{\tau-\rho}{\omega n}} d\rho + \mathcal{R}_1 \\ &= x_\omega^* (y(\tau_0)) \left(1 - e^{-\frac{\tau-\tau_0+\delta_1}{\omega n}}\right) + \mathcal{R}_1, \end{aligned}$$

where

$$\mathcal{R}_1 = \frac{1}{\omega n} \int_{\tau_0-\delta_1}^{\tau} e^{-\frac{\tau-\rho}{\omega n}} x_\omega^* (y(\rho) - y(\tau_0)) \, d\rho.$$

Since $(\tau_0 - \delta_1, \tau) \subset (\tau_0 - \delta_1, \tau_0 + \eta) \subset (\tau_0 - \delta_1, \tau_0 + \delta_1)$, it follows from the definition of δ_1 that

$$| \mathcal{R}_1 | < \frac{\varepsilon}{4} \cdot \frac{1}{\omega n} \int_{\tau_0-\delta_1}^{\tau} e^{-\frac{\tau-\rho}{\omega n}} d\rho < \frac{\varepsilon}{4}.$$

Similarly,

$$\frac{1}{\omega n} \int_{\tau_0-\delta_1}^{\tau_0} e^{-\frac{\tau_0-\rho}{\omega n}} x_\omega^* (y(\rho)) \, d\rho = x_\omega^* (y(\tau_0)) \left(1 - e^{-\frac{\delta_1}{\omega n}}\right) + \mathcal{R}_2,$$

$$| \mathcal{R}_2 | < \frac{\varepsilon}{4}.$$

The latter estimates imply the inequality

$$\Delta_2 \leqslant \| y(\tau_0) \|_{C(\Omega)} \left| e^{-\frac{\tau-\tau_0+\delta_1}{\omega n}} - e^{-\frac{\delta_1}{\omega n}} \right| + \frac{\varepsilon}{2}.$$

The difference between the exponentials can be estimated by means of the device used above. This yields

$$\Delta_2 \leqslant \frac{3\eta}{2e\delta_1} \| y(\tau_0) \|_{C(\Omega)} + \frac{\varepsilon}{2}$$

and

$$| z(\tau, \omega) - z(\tau_0, \omega) | \leqslant \frac{\varepsilon}{2} + \eta \left\{ \frac{9}{e^2 \delta_1^2} \int_0^{\tau_0} \| y(\rho) \|_{C(\Omega)} \, d\rho + \frac{3}{2e\delta_1} \| y(\tau_0) \|_{C(\Omega)} \right\} < \varepsilon$$

for $\omega n > 0$.

Let us consider the case $\omega n < 0$. We have

$$| z(\tau, \omega) - z(\tau_0, \omega) | \leqslant \Delta_3 + \Delta_4,$$

$$\Delta_3 = \left| \frac{1}{|\omega n|} \int_{\tau_0+\delta_1}^{\infty} \left(e^{-\frac{\rho-\tau}{|\omega n|}} - e^{-\frac{\rho-\tau_0}{|\omega n|}} \right) x_\omega^* (y(\rho)) \, d\rho \right|;$$

$$\Delta_4 = \left| \frac{1}{|\omega n|} \int_{\tau}^{\tau_0+\delta_1} e^{-\frac{\rho-\tau}{|\omega n|}} x_\omega^* (y(\rho)) \, d\rho - \frac{1}{|\omega n|} \int_{\tau_0}^{\tau_0+\delta_1} e^{-\frac{\rho-\tau_0}{|\omega n|}} x_\omega^* (y(\rho)) \, d\rho \right|.$$

$$\Delta_3 \leqslant \frac{3}{\delta_1} \left[\frac{\delta_1}{3 |\omega n|} e^{-\frac{\delta_1}{3|\omega n|}} \right] \int_{\tau_0+\delta_1}^{\infty} \| y(\rho) \|_{C(\Omega)} e^{-\frac{\rho-\tau_0-\delta_1}{|\omega n|}} \left| e^{-\frac{\frac{2\delta_1}{3} - (\tau-\tau_0)}{|\omega n|}} - e^{-\frac{2\delta_1}{3|\omega n|}} \right| d\rho$$

$$\leqslant \frac{3}{\delta_1} \cdot \frac{1}{e} \int_{\tau_0+\delta_1}^{\infty} \| y(\rho) \|_{C(\Omega)} e^{-\rho+\tau_0+\delta_1} \cdot \frac{3\eta}{e\delta_1} \, d\rho \leqslant \frac{9\eta}{e^2\delta_1^2} e^{\tau_0+\delta_1} \int_{\tau_0+\delta_1}^{\infty} e^{-\rho} \| y(\rho) \|_{C(\Omega)} \, d\rho.$$

Further,

$$\frac{1}{|\omega n|} \int_{\tau}^{\tau_0+\delta_1} e^{-\frac{\rho-\tau}{|\omega n|}} x_\omega^* (y(\rho)) \, d\rho = x_\omega^* (y(\tau_0)) \left(1 - e^{-\frac{\tau_0-\tau+\delta_1}{|\omega n|}} \right) + \mathscr{R}_3,$$

$$\frac{1}{|\omega n|} \int_{\tau_0}^{\tau_0+\delta_1} e^{-\frac{\rho-\tau_0}{|\omega n|}} x_\omega^* (y(\rho)) \, d\rho = x_\omega^* (y(\tau_0)) \left(1 - e^{-\frac{\delta_1}{|\omega n|}} \right) + \mathscr{R}_4,$$

$$| \mathscr{R}_3 | < \frac{\varepsilon}{4}, \quad | \mathscr{R}_4 | < \frac{\varepsilon}{4}.$$

Hence

$$\Delta_4 < \frac{\varepsilon}{2} + \| y(\tau_0) \|_{C(\Omega)} \left| e^{-\frac{\tau_0-\tau+\delta_1}{|\omega n|}} - e^{-\frac{\delta_1'}{|\omega n|}} \right| \leqslant \frac{\varepsilon}{2} + \frac{3\eta}{2e\delta_1} \| y(\tau_0) \|_{C(\Omega)}.$$

Thus, if $\omega n < 0$, then for $|\tau - \tau_0| < \eta$ we have

$$|z(\tau, \omega) - z(\tau_0, \omega)| \leqslant \frac{\varepsilon}{2}$$

$$+ \eta \left[\frac{9}{e^2 \delta_1^2} e^{\tau_0 + \delta_1} \int_{\tau_0 + \delta_1}^{\infty} e^{-\rho} \| y(\rho) \|_{C(\Omega)} d\rho + \frac{3}{2e\delta_1} \| y(\tau_0) \|_{C(\Omega)} \right] < \varepsilon.$$

Finally, if $\omega n = 0$, then for $|\tau - \tau_0| < \eta < \delta_1$ we have

$$|z(\tau, \omega) - z(\tau_0, \omega)| = |x^*_\omega (y(\tau)) - x^*_\omega (y(\tau_0))| \leqslant \| y(\tau) - y(\tau_0) \|_{C(\Omega)} \leqslant \frac{\varepsilon}{4} < \varepsilon.$$

Thus

$$|z(\tau, \omega) - z(\tau_0, \omega)| < \varepsilon,$$

for $|\tau - \tau_0| < \eta$ for every $\omega \in \Omega$. The theorem is proved.

Corollary. *Let* $x(\cdot) \in \mathcal{B}_1$, $0 \leq \tau_1 < \tau_2 < \infty$, *and let* $x(\tau)$ *be strongly continuous with respect to* τ *in the interval* (τ_1, τ_2). *Let* $z(\cdot) = \hat{A}x(\cdot)$. *The numerical function* $z(\tau, \omega)$ *is then continuous with respect to the pair of variables* (τ, ω) *on the set* $(\tau_1, \tau_2) \times \Omega$.

That Theorem 17 [vi] implies this statement is self-evident.

All of the facts set forth in Theorem 17 concern the behavior of the functions $\hat{A}x(\tau)$, $x(\cdot) \in \mathcal{B}_1$, for positive values of τ. However, formulas (57)–(59) enable us to determine $\hat{A}x(\tau)$ for $\tau = 0$ as well. These boundary values of the functions $Ax(\cdot)$ turn out to be related to the values of $\hat{A}x(\tau)$ for $\tau > 0$ by natural limit relations.

For $x(\cdot) \in \mathcal{B}_1$ we set

$$z^0(\omega) = \begin{cases} \frac{1}{|\omega n|} \int_0^\infty e^{-\rho/|\omega n|} x^*_\omega (\hat{g}x(\rho)) \, d\rho & \text{for } \omega n < 0, \\ 0 & \text{for } \omega n \geqslant 0. \end{cases} \tag{63}$$

As above (compare formulas (57) and (58)), the inclusion $x(\cdot) \in \mathcal{B}_1$ ensures the single-valued definiteness and finiteness of $z^0(\omega)$ for every $\omega \in \Omega$.

Theorem 18. *Let* $x(\cdot) \in \mathcal{B}_1$, $z(\cdot) = \hat{A}x(\cdot)$, *and let* z^0 *be defined in accordance with* (63). *Then*

[i] $\sup_{\omega \in \Omega} |z^0(\omega)| < \infty$.

[ii] *The function* $z^0(\omega)$ *is continuous at any point* $\omega_0 \in \Omega$ *for which* $\omega_0 n \neq 0$.

[iii] $z^0 \in L_2(\Omega)$.

[iv] $\lim_{\tau = 0+0} z(\tau, \omega) = z^0(\omega)$ *uniformly for every* $\delta > 0$ *with respect to all*

$\omega \in \Omega$ *satisfying the condition* $|\omega\mathbf{n}| \geq \delta$.

[v] $\lim_{\tau=0+0} \|z(\tau) - z^0\| = 0$.

The proof of this theorem is very similar to the proof of Lemma 20. In fact, statement [i] of the theorem can be proved in exactly the same way as the boundedness of $z_-(\tau, \omega)$. Repeating the approximate operations, we obtain

$$|z^0(\omega)| \leqslant \sqrt{2\pi}\,[g]\left\{l_0(1) + \int_1^\infty e^{-\rho}\|x(\rho)\|\,d\rho\right\} < \infty, \quad \omega \in \Omega.$$

To prove [ii] we note that $z^0(\omega)$ is (self-evidently) continuous at the point ω_0 if $\omega_0\mathbf{n} > 0$. On the other hand, if $\omega_0\mathbf{n} < 0$, then the continuity of $z^0(\omega)$ at the point ω_0 can be established in the same way as the continuity of $z_-(\tau, \omega)$ with respect to ω in the proof of Lemma 20.

Statement [iii] of the theorem follows directly from [i] and [ii]. If $\omega\mathbf{n} > \delta > 0$, then for $0 < \tau < 1$ we have

$$|z(\tau, \omega) - z^0(\omega)| \leqslant \sqrt{2\pi}\,[g]\,l_0(1)\,(1 - e^{-\tau/\delta}).$$

If, on the other hand, $\omega\mathbf{n} < -\delta$, then for $0 < \tau < 1$

$$|z(\tau, \omega) - z^0(\omega)| \leqslant 2\sqrt{2\pi}\,[g]\,l_0(3)\,(1 - e^{-\tau/\delta}) + 2\pi\,[g]\,(e^\tau - 1)\int_3^\infty e^{-\rho}\|x(\rho)\|\,d\rho.$$

These relations follow if we replace τ by 0 and τ' by τ in those relations in the proof of Lemma 20 which yield the estimate of the difference $|z_\pm(\tau, \omega) - z_\pm(\tau', \omega)|$ and then repeat the argument. Statement [iv] of the theorem now becomes self-evident.

Finally, we note that $\|z(\tau) - z^0\|^2 = \int_\Omega |z(\tau, \omega) - z^0(\omega)|^2\,d\omega$. According to [iv], $\lim_{\tau=0+0} |z(\tau, \omega) - z^0(\omega)| = 0$ for every $\omega \in \Omega$ for which $\omega\mathbf{n} \neq 0$. On the other hand,

$$|z(\tau, \omega) - z^0(\omega)| \leqslant \sup_{\omega\in\Omega} |z^0(\omega)| + l(1) = \text{const} < \infty$$

immediately for all $(\tau, \omega) \in (0, 1] \times \Omega$ (compare Theorem 17 [ii]). Making use of the Lebesgue theorem on majorized convergence, we arrive at [v]. The theorem is proved.

Theorem 19. *Let* $x(\cdot) \in \mathfrak{B}_1$, $z(\cdot) = \hat{A}x(\cdot)$, *and let the limit* $\lim_{\tau=0+0} x(\tau) = x_0 \in L_2(\Omega)$ *exist (in the metric of* $L_2(\Omega)$). *We set*

$$z^{00}(\omega) = \begin{cases} z^0(\omega) & \text{for } \omega\mathbf{n} \neq 0. \\ x^*_\omega(\hat{g}x_0) & \text{for } \omega\mathbf{n} = 0. \end{cases}$$

Then

[i] $z^{00}(\omega)$ *is continuous on the closed hemisphere*

$$\Omega^- = \{\omega \mid \omega \in \Omega,\ \omega\mathbf{n} \leqslant 0\}.$$

[ii] $\lim_{\tau=0+0} z(\tau, \omega) = z^{00}(\omega)$ *uniformly with respect to all* $\omega \in \Omega^-$.

Proof. Theorem 18 implies that $z^{00}(\omega)$ is bounded on Ω and continuous at every point $\omega_0 \in \Omega^-$ for which $\omega_0\mathbf{n} \neq 0$. We set $y(\tau) = \hat{g}x(\tau)$, $\tau > 0$, and $y_0 = \hat{g}x_0$. Then $y_0 \in C(\Omega)$, $y(\tau) \in C(\Omega)$, $\tau > 0$ and $\lim_{\tau=0+0} \|y(\tau) - y_0\|_{C(\Omega)} = 0$.

Let $\epsilon > 0$. We choose $\delta_1 > 0$ in such a way that $\|y(\tau) - y_0\|_{C(\Omega)} < \epsilon/4$ for $0 < \tau < \delta_1$. We now choose a $\delta_2 > 0$ such that

a) for $\omega_{1,2} \in \Omega$, $\omega_1 - \omega_2 < \delta_2$,

$$|y_0(\omega_1) - y_0(\omega_2)| < \frac{\varepsilon}{4};$$

b) for $0 < \mu < \delta_2$,

$$\frac{1}{\mu} e^{-\delta_1/\mu} < \frac{\varepsilon}{4}\left\{1 + e^{\delta_1}\int_{\delta_1}^{\infty} e^{-\rho}\left(\|y(\rho)\|_{C(\Omega)} + \|y_0\|_{C(\Omega)}\right) d\rho\right\}^{-1}.$$

Let us choose an $\omega_0 \in \Omega^-$, $\omega_0\mathbf{n} = 0$, and let $\omega \in \Omega^-$, $\omega - \omega_0 < \delta_2$. If $\omega\mathbf{n} < 0$, then arguments entirely similar to those used in proving Theorem 17 [vi] yield the inequality

$$|z^{00}(\omega) - z^{00}(\omega_0)| < \varepsilon.$$

If, on the other hand, $\omega\mathbf{n} = 0$, then

$$|z^{00}(\omega) - z^{00}(\omega_0)| = |y_0(\omega) - y_0(\omega_0)| < \frac{\varepsilon}{4} < \varepsilon.$$

Thus $z^{00}(\omega)$ is continuous on Ω^-. To prove the limit relation [ii] we proceed on the basis of the $\delta_1 = \delta_1(\epsilon)$ chosen above and select the number η in such a way that

$$0 < \eta \leqslant \min\left\{\frac{\delta_1}{3},\ \frac{\varepsilon}{3}\left[1 + \frac{9}{\delta_1^2 e^2} e^{\delta_1}\int_{\delta_1}^{\infty} e^{-\rho}\|y(\rho)\|_{C(\Omega)} d\rho + \frac{3}{2e\delta_1}\|y_0\|_{C(\Omega)}\right]^{-1}\right\}.$$

Let $0 < \tau < \eta$ and $\omega \in \Omega^-$. If $\omega\mathbf{n} < 0$, then, making use of the devices used to prove Theorem 17 [vi], we obtain

$$|z(\tau, \omega) - z^{00}(\omega)| < \varepsilon.$$

On the other hand, if $\omega\mathbf{n} = 0$, then

$$|z(\tau, \omega) - z^{00}(\omega)| = |x^*_\omega(y(\tau)) - x^*_\omega(y_0)| \leqslant \|y(\tau) - y_0\|_{C(\Omega)} < \frac{\varepsilon}{4} < \varepsilon.$$

This means that $z(\tau, \omega)$ converges uniformly to $z^{00}(\omega)$ with respect to $\omega \in \Omega^-$ as $\tau \to 0 + 0$.

Let us note yet another important property of the operator $\hat{A}$.

Lemma 21. *Let* $x(\cdot) \in \mathfrak{B}_1$, $\hat{A}x(\cdot) = z(\cdot)$, *and assume that* $x(\tau) = \hat{P}^{(0)}_{\mathbf{n}} x(\tau)$ *for every* $\tau \in (0, \infty)$. *Then it is also the case that* $z(\tau) = \hat{P}^{(0)}_{\mathbf{n}} z(\tau)$ *for every* $\tau \in (0, \infty)$.

Proof. Let $y(\tau) = \hat{g}x(\tau)$, $\tau > 0$. Since g commutes with $\hat{P}^{(0)}_{\mathbf{n}}$ (Theorem 2), we have $\hat{P}^{(0)}_{\mathbf{n}} y(\tau) = y(\tau)$, $\tau > 0$. According to Lemma 13, to every $\tau > 0$ we can assign a function $a_\tau(\mu)$ continuous on $[-1, 1]$ and such that $a_\tau(\omega\mathbf{n}) = y(\tau, \omega) = x^*_\omega(y(\tau))$, $\omega \in \Omega$. Making use of the definition of $\hat{A}$ and formulas (57)–(59), we find that $z(\tau, \omega_1) = z(\tau, \omega_2)$ in all cases when $\omega_1\mathbf{n} = \omega_2\mathbf{n}$, $\tau > 0$.

Let us choose an arbitrary orthonormal basis σ of the space R_3 with the third unit vector $\mathbf{n}$. According to what we said above, for any $\mu \in [-1, 1]$ we have

$$z(\tau, \psi_\sigma(\mu, \varphi_1)) = z(\tau, \psi_\sigma(\mu, \varphi_2)),$$

regardless of what the $\phi_{1,2} \in [0, 2\pi]$ happen to be. This means that there exists a function $b_\tau(\mu)$, $\tau > 0$, defined for $\mu \in [-1, 1]$, such that $z(\tau, \psi_\sigma(\mu, \phi)) = b_\tau(\mu)$ for all $(\mu, \phi) \in \Delta$.

Let $n \geq 0$, $-n \leq m \leq n$, $m \neq 0$. Then (compare Lemmas 1 and 2)

$$(z(\tau), Y^{(m)}_{\sigma n}) = \int_\Delta z(\tau, \psi_\sigma(\mu, \varphi)) \overline{Y^{(m)}_n(\mu, \varphi)}\, d\mu\, d\varphi$$

$$= \int_{-1}^{1} b_\tau(\mu)\, d\mu \int_0^{2\pi} \overline{Y^{(m)}_n(\mu, \varphi)}\, d\varphi = 0.$$

Now, applying Lemmas 2[ii] and 3 for $\tau > 0$, we obtain

$$z(\tau) = \sum_{n=0}^{\infty} \sum_{m=-n}^{n} (z(\tau), Y^{(m)}_{\sigma n}) Y^{(m)}_{\sigma n} = \sum_{n=0}^{\infty} (z(\tau), Y^{(0)}_{\sigma n}) Y^{(0)}_{\sigma n} = \hat{P}^{(0)}_{\mathbf{n}} z(\tau),$$

Q.E.D.

§8. Invariance of the classes $\mathfrak{B}_\epsilon$

It is natural to attempt to construct solutions of the homogeneous and inhomogeneous Milne problems by the method of successive approximations. In applying this method it is generally important to ensure that the class of functions in which the solution is sought is invariant with respect to the operator $\hat{A}$. As we shall see below (Theorem 22), the class $\mathfrak{B}_1$ does not have this property. On the other hand, all of the $\mathfrak{B}_\epsilon$, $-1 < \epsilon < 1$, are invariant with respect to $\hat{A}$ (Theorem 20).

In the preceding section we showed that the operator $\hat{A}$ improves the local properties of functions considerably. This is especially true of high iterations of $\hat{A}$. In fact (compare Theorem 17), if $x(\cdot) \in \mathfrak{B}_\epsilon$, $|\epsilon| < 1$, and if $x_n(\cdot) = \hat{A}^n(x)$, then $x_n(\tau)$, $n \geq 1$, is a strongly continuous function of τ, $x_n(\cdot) \in \mathfrak{B}_\epsilon$. This in turn implies that the numerical functions $x_n(\tau, \omega)$ are continuous with respect to the pair of variables $(\tau, \omega) \in (0, \infty) \times \Omega$ for $n \geq 2$. It turns out that the $x_n(\tau, \omega)$, $n \geq 4$, admit of a convenient majorant which does not depend on ω (Theorem 21). This makes it much easier for us to prove the convergence of the successive approximations.

By definition (§7) the operator $\hat{A} = \hat{A}_g$ depends on the choice of the function g. As in the previous section, most of the results developed in the present section are valid under the sole restriction $g \in L_2(-1, 1)$. Only in Theorem 22 do we impose stronger restrictions on the function g.

Theorem 20. *Let $g \in L_2(-1, 1)$ and $\hat{A} = \hat{A}_g$. Then for $\epsilon \in (-1, 1)$ we have*

$$\hat{A}(\mathfrak{B}_\epsilon) \subset \mathfrak{B}_\epsilon.$$

Proof. Let $x(\cdot) \in \mathfrak{B}_\epsilon$, $|\epsilon| < 1$. We set $z(\cdot) = \hat{A}x(\cdot)$. Then $z(\cdot) \in \mathfrak{K}$ and $z(\tau)$ is a strongly continuous function of τ with values in $L_2(\Omega)$ (Theorem 17[i]). This means, among other things, that $z(\tau)$ is strongly measurable with respect to τ on $(0, \infty)$. $\|z(\tau)\|$ is bounded on every bounded portion of $(0, \infty)$ (Theorem 17[ii]). To prove this theorem it is clearly sufficient to prove that the integral $\int_0^\infty \|z(\tau)\| e^{-\epsilon\tau} d\tau$ is finite. We have

$$\|z(\tau)\| \leqslant \|z_-(\tau)\| + \|z_+(\tau)\|,$$

where (Lemma 20[vi])

$$\int_0^\infty e^{-\varepsilon\tau}\|z_-(\tau)\|d\tau \leqslant \pi\sqrt{2}\,[g]\int_0^\infty e^{(1-\varepsilon)\tau}d\tau\int_\tau^\infty e^{-\rho}\frac{\|x(\rho)\|}{\sqrt{\rho-\tau}}d\rho$$

$$= \pi\sqrt{2}\,[g]\int_0^\infty e^{-\rho}\|x(\rho)\|d\rho\int_0^\rho\frac{e^{(1-\varepsilon)\tau}}{\sqrt{\rho-\tau}}d\tau.$$

But

$$e^{-(1-\varepsilon)\rho}\int_0^\rho\frac{e^{(1-\varepsilon)\tau}}{\sqrt{\rho-\tau}}d\tau = \int_0^\rho\frac{e^{-(1-\varepsilon)\xi}}{\sqrt{\xi}}d\xi < \int_0^\infty \xi^{-1/2}e^{-(1-\varepsilon)\xi}d\xi = \sqrt{\frac{\pi}{1-\varepsilon}},$$

so that

$$\int_0^\infty e^{-\varepsilon\tau}\|z_-(\tau)\|d\tau \leqslant \sqrt{\frac{2\pi^3}{1-\varepsilon}}\,[g]\int_0^\infty e^{-\varepsilon\rho}\|x(\rho)\|d\rho < \infty.$$

Similarly,

$$\int_0^\infty e^{-\varepsilon\tau}\|z_+(\tau)\|d\tau \leqslant \pi\sqrt{2}\,[g]\int_0^\infty e^{-\varepsilon\tau}d\tau\int_0^\tau e^{-(\tau-\rho)}\frac{\|x(\rho)\|}{\sqrt{\tau-\rho}}d\rho$$

$$= \pi\sqrt{2}\,[g]\int_0^\infty e^{\rho}\|x(\rho)\|d\rho\int_\rho^\infty e^{-(1+\varepsilon)\tau}\frac{d\tau}{\sqrt{\tau-\rho}} = \sqrt{\frac{2\pi^3}{1+\varepsilon}}\,[g]\int_0^\infty e^{-\varepsilon\rho}\|x(\rho)\|d\rho < \infty.$$

The theorem is proved.

Theorem 21. *Let* $g \in L_2(-1, 1)$, $\hat{A} = \hat{A}_g$. *If* $\epsilon \in (-1, 1)$, $x(\cdot) \in \mathfrak{B}_\epsilon$, *then*

$$k_0 \equiv \sup_{\tau>0} e^{-\varepsilon\tau} \| \hat{A}^3 x(\tau) \| < \infty,$$

and for all $(\tau, \omega) \in (0, \infty) \times \Omega$ *we have*

$$|\hat{A}^4 x(\tau, \omega)| \leqslant \frac{\sqrt{2\pi}\,[g]}{1-|\varepsilon|} k_0 e^{\varepsilon\tau}.$$

Proof. Let $\epsilon \in (-1, 1)$. We denote by $\mathfrak{U}$ the collection of all nonnegative numerical functions summable on $(0, \infty)$, each of which is bounded on any bounded portion of $(0, \infty)$. We introduce the operators $\hat{X}$ and $\hat{Y}$ acting on the functions from $\mathfrak{U}$ according to the formulas

$$\hat{X}f(\tau) = e^{(1-\varepsilon)\tau} \int_\tau^\infty e^{-(1-\varepsilon)\rho} \frac{f(\rho)}{\sqrt{\rho-\tau}} d\rho,$$

$$\hat{Y}f(\tau) = e^{-(1+\varepsilon)\tau} \int_0^\tau e^{(1+\varepsilon)\rho} \frac{f(\rho)}{\sqrt{\tau-\rho}} d\rho, \quad f \in \mathfrak{U}, \ \tau > 0.$$

Let us verify that $\mathfrak{U}$ is invariant with respect to $\hat{X}$ and $\hat{Y}$. In fact, $\hat{X}f(\tau)$ and $\hat{Y}f(\tau)$ for $f \in \mathfrak{U}$ are nonnegative functions measurable on $(0, \infty)$ (the Fubini theorem). If $\tau_1 > 0$, then for $\tau \in (0, \tau_1]$ we have

$$\hat{X}f(\tau) \leqslant \int_\tau^{\tau_1+1} \frac{f(\rho)}{\sqrt{\rho-\tau}} d\rho + \int_{\tau_1+1}^\infty \frac{f(\rho)}{\sqrt{\rho-\tau}} d\rho$$

$$\leqslant 2\sqrt{1+\tau_1} \sup_{\rho\in(0,\tau_1+1]} f(\rho) + \int_{\tau_1+1}^\infty f(\rho)\,d\rho < \infty;$$

$$\hat{Y}f(\tau) \leqslant 2\sqrt{\tau_1} \sup_{\rho\in(0,\tau_1]} f(\rho) < \infty,$$

so that $\hat{X}f(\tau)$ and $\hat{Y}f(\tau)$ are bounded on every bounded portion of $(0, \infty)$. Finally,

$$\int_0^\infty \hat{X}f(\tau)\,d\tau = \int_0^\infty f(\rho)\, e^{-(1-\varepsilon)\rho} d\rho \int_0^\rho \frac{e^{(1-\varepsilon)\tau}}{\sqrt{\rho-\tau}} d\tau \leqslant \sqrt{\frac{\pi}{1-\varepsilon}} \int_0^\infty f(\rho)\,d\rho < \infty;$$

$$\int_0^\infty \hat{Y}f(\tau)\,d\tau = \int_0^\infty e^{(1+\varepsilon)\rho} f(\rho)\,d\rho \int_\rho^\infty \frac{e^{-(1+\varepsilon)\tau}}{\sqrt{\tau-\rho}} d\tau = \sqrt{\frac{\pi}{1+\varepsilon}} \int_0^\infty f(\rho)\,d\rho < \infty.$$

Thus $\hat{X}(\mathfrak{U}) \subset \mathfrak{U}$, $\hat{Y}(\mathfrak{U}) \subset \mathfrak{U}$.

Now let $x(\cdot) \in \mathfrak{B}_\epsilon$ and $x_n(\cdot) = \hat{A}^n x(\cdot)$, $n = 1, 2, \cdots$. We set

$$f_n(\tau) = e^{-\varepsilon\tau} \|x_n(\tau)\|, \ \tau > 0, \ n = 1, 2, \ldots; f(\tau) = e^{-\varepsilon\tau} \|x(\tau)\|.$$

As we see from Theorem 20, $f_n \in \mathfrak{U}$, $n \geq 1$. It is clear that $f \in \mathfrak{U}$ as well. By

Lemma 20 [vi] we have

$$f_1(\tau) \leqslant \pi \sqrt{2}\,[g]\,e^{-\varepsilon\tau}\left(\int_0^\tau e^{-(\tau-\rho)}\frac{1}{\sqrt{\tau-\rho}}\,e^{\varepsilon\rho} f(\rho)\,d\rho\right.$$

$$\left.+\int_\tau^\infty e^{-(\rho-\tau)}\frac{1}{\sqrt{\rho-\tau}}\,e^{\varepsilon\rho} f(\rho)\,d\rho\right) = \pi\sqrt{2}\,[g]\,(\hat{X}+\hat{Y})\,f(\tau).$$

Similarly,

$$\hat{X}^3 f(\tau),\ \hat{X}^2\hat{Y} f(\tau),\ \hat{X}\hat{Y}\hat{X} f(\tau),\ \hat{X}\hat{Y}^2 f(\tau),\ \hat{Y}\hat{X}^2 f(\tau),\ \hat{Y}\hat{X}\hat{Y} f(\tau),\ \hat{Y}^2\hat{X} f(\tau),\ \hat{Y}^3 f(\tau)$$

In order to prove the first statement of the theorem, we must clearly verify that $\sup_{\tau>0} f_3(\tau) < \infty$. To do this we need merely prove the boundedness on $(0, \infty)$ of the eight functions

$$f_n(\tau) \leqslant (\pi\sqrt{2}\,[g])^n\,(\hat{X}+\hat{Y})^n f(\tau),\quad \tau > 0,\ n = 1, 2, \ldots$$

for $f \in \mathfrak{U}$.

We begin by estimating the quadratic expressions

$$\hat{X}^2 f(\tau),\ \hat{X}\hat{Y} f(\tau),\ \hat{Y}\hat{X} f(\tau),\ \hat{Y}^2 f(\tau),\ f \in \mathfrak{U}.$$

Simple computation shows that

$$\hat{X}^2 f(\tau) = \pi e^{(1-\varepsilon)\tau}\int_\tau^\infty e^{-(1-\varepsilon)\rho} f(\rho)\,d\rho, \tag{64}$$

$$\hat{Y}^2 f(\tau) = \pi e^{-(1+\varepsilon)\tau}\int_0^\tau e^{(1+\varepsilon)\rho} f(\rho)\,dp. \tag{65}$$

We set

$$a = \ln(1+\sqrt{2}) + \int_1^\infty e^{-2s^2}\,ds;$$

$$h(x) = a + \begin{cases} \frac{1}{2}\ln\frac{1}{x} & \text{for } x \in (0, 1), \\ 0 & \text{for } x \geqslant 1, \end{cases}$$

$$\hat{Z} f(\tau) = 2e^{-(1+\varepsilon)\tau}\int_0^\tau e^{(1+\varepsilon)\rho} h(\tau-\rho) f(\rho)\,d\rho +$$

$$+ 2e^{(1-\varepsilon)\tau}\int_\tau^\infty e^{-(1-\varepsilon)\rho} h(\rho-\tau) f(\rho)\,d\rho,\ f \in \mathfrak{U}.$$

We obtain

$$\hat{X}\hat{Y} f(\tau) = e^{(1-\varepsilon)\tau}\int_0^\tau e^{(1+\varepsilon)\rho} f(\rho)\,d\rho\int_\tau^\infty \frac{e^{-2\tau'}}{\sqrt{\tau'-\tau}\,\sqrt{\tau'-\rho}}\,d\tau' +$$

$$+ e^{(1-\varepsilon)\tau} \int_\tau^\infty e^{(1+\varepsilon)\rho} f(\rho)\, d\rho \int_\rho^\infty \frac{e^{-2\tau'}}{\sqrt{\tau'-\tau}\,\sqrt{\tau'-\rho}}\, d\tau'.$$

For $\rho < \tau$ we have

$$\int_\tau^\infty \frac{e^{-2\tau'}}{\sqrt{\tau'-\tau}\,\sqrt{\tau'-\rho}}\, d\tau' = 2e^{-2\tau} \int_0^\infty \frac{e^{-2s^2}\, ds}{\sqrt{s^2+(\tau-\rho)}}$$

$$\leqslant 2e^{-2\tau} \left\{ \int_0^1 \frac{ds}{\sqrt{s^2+(\tau-\rho)}} + \int_1^\infty e^{-2s^2}\, ds \right\}.$$

If $\tau - \rho \geq 1$, then

$$\int_0^1 \frac{ds}{\sqrt{s^2+(\tau-\rho)}} \leqslant \int_0^1 \frac{ds}{\sqrt{s^2+1}} = \ln(1+\sqrt{2}).$$

If $\tau - \rho < 1$, then

$$\int_0^1 \frac{ds}{\sqrt{s^2+(\tau-\rho)}} = \int_0^{\sqrt{\tau-\rho}} \frac{ds}{\sqrt{s^2+(\tau-\rho)}} + \int_{\sqrt{\tau-\rho}}^1 \frac{ds}{\sqrt{s^2+(\tau-\rho)}} < \ln(1+\sqrt{2}) + \frac{1}{2}\ln\frac{1}{\tau-\rho}.$$

Hence for $\rho < \tau$ we obtain

$$\int_\tau^\infty \frac{e^{-2\tau'}}{\sqrt{\tau'-\tau}\,\sqrt{\tau'-\rho}}\, d\tau' \leqslant 2e^{-2\tau} h(\tau-\rho).$$

Interchanging ρ and τ, we find that for $\tau < \rho$

$$\int_\rho^\infty \frac{e^{-2\tau'}\, d\tau'}{\sqrt{\tau'-\tau}\,\sqrt{\tau'-\rho}} \leqslant 2e^{-2\rho} h(\rho-\tau).$$

We can now write

$$\hat{X}\hat{Y} f(\tau) \leqslant 2e^{-(1+\varepsilon)\tau} \int_0^\tau e^{(1+\varepsilon)\rho} h(\tau-\rho) f(\rho)\, d\rho$$

$$+ 2e^{(1-\varepsilon)\tau} \int_\tau^\infty e^{-(1-\varepsilon)\rho} h(\rho-\tau) f(\rho)\, d\rho,$$

or, which is the same thing,

$$\hat{X}\hat{Y} f(\tau) \leqslant \hat{Z} f(\tau). \tag{66}$$

Similar constructions yield the inequality

$$\hat{Y}\hat{X} f(\tau) \leqslant \hat{Z} f(\tau). \tag{67}$$

Let us now estimate the cubic expressions. We have

$$\hat{X}^3 f(\tau) = \pi e^{(1-\varepsilon)\tau} \int_\tau^\infty e^{-(1-\varepsilon)\rho} f(\rho)\, d\rho \int_\tau^\rho \frac{d\tau'}{\sqrt{\rho-\tau'}} =$$

$$= \frac{2\pi}{\sqrt{1-\varepsilon}} \int_{\tau}^{\infty} [\sqrt{(1-\varepsilon)(\rho-\tau)}\, e^{-(1-\varepsilon)(\rho-\tau)}]\, f(\rho)\, d\rho,$$

from which we obtain

$$\hat{X}^3 f(\tau) \leqslant \pi \sqrt{\frac{2}{e(1-\varepsilon)}} \int_{\tau}^{\infty} f(\rho)\, d\rho. \tag{68}$$

Similarly,

$$\hat{Y}^3 f(\tau) \leqslant \pi \sqrt{\frac{2}{e(1+\varepsilon)}} \int_{0}^{\tau} f(\rho)\, d\rho. \tag{69}$$

$$\hat{X}^2\hat{Y} f(\tau) = \pi e^{(1-\varepsilon)\tau} \int_0^{\tau} e^{(1+\varepsilon)\rho} f(\rho)\, d\rho \int_{\tau}^{\infty} e^{-2\tau'} \frac{d\tau'}{\sqrt{\tau'-\rho}}$$

$$+ \pi e^{(1-\varepsilon)\tau} \int_{\tau}^{\infty} e^{(1+\varepsilon)\rho} f(\rho)\, d\rho \int_{\rho}^{\infty} e^{-2\tau'} \frac{d\tau'}{\sqrt{\tau'-\rho}} \leqslant \pi e^{-(1+\varepsilon)\tau} \int_0^{\tau} e^{(1+\varepsilon)\rho} f(\rho)\, d\rho \int_0^{\infty} e^{-2\tau'} \frac{d\tau'}{\sqrt{\tau'}}$$

$$+ \pi e^{(1-\varepsilon)\tau} \int_{\tau}^{\infty} e^{-(1-\varepsilon)\rho} f(\rho)\, d\rho \int_0^{\infty} e^{-2\tau'} \frac{d\tau'}{\sqrt{\tau'}} \leqslant \sqrt{\frac{\pi^3}{2}} \int_0^{\infty} f(\rho)\, d\rho,$$

$$\hat{X}^2\hat{Y} f(\tau) \leqslant \sqrt{\frac{\pi^3}{2}} \int_0^{\infty} f(\rho)\, d\rho. \tag{70}$$

Completely analogous operations yield the estimates

$$\hat{X}\hat{Y}^2 f(\tau) \leqslant \sqrt{\frac{\pi^3}{2}} \int_0^{\infty} f(\rho)\, d\rho, \tag{71}$$

$$\hat{Y}\hat{X}^2 f(\tau) \leqslant \sqrt{\frac{\pi^3}{2}} \int_0^{\infty} f(\rho)\, d\rho, \tag{72}$$

$$\hat{Y}^2\hat{X} f(\tau) \leqslant \sqrt{\frac{\pi^3}{2}} \int_0^{\infty} f(\rho)\, d\rho. \tag{73}$$

To estimate the expression $\hat{X}\hat{Y}\hat{X}f(\tau)$ we note, to begin with, that

$$\int_0^{\infty} e^{-2x^2} h(x^2)\, dx < 1 + \frac{a}{2}\sqrt{\frac{\pi}{2}},$$

and that for $p > 0$ we have

$$\int_0^1 h(p(1-y^2))\, dy \leqslant a_1 - \begin{cases} \frac{1}{2}\ln p, & \text{if } p \leqslant 1, \\ 0 & \text{if } p > 1, \end{cases}$$

where $a_1 = a + 1 - \ln 2$.

Now we see that

$$\hat{X}\hat{Y}\hat{X}f(\tau) \leqslant \hat{X}\hat{Z}f(\tau) = 2e^{(1-\varepsilon)\tau}\int_0^\tau e^{(1+\varepsilon)\rho} f(\rho)\,d\rho \int_\tau^\infty \frac{e^{-2\tau'} h(\tau'-\rho)}{\sqrt{\tau'-\tau}}\,d\tau'$$

$$+ 2e^{(1-\varepsilon)\tau}\int_\tau^\infty e^{(1+\varepsilon)\rho} f(\rho)\,d\rho \int_\rho^\infty \frac{e^{-2\tau'} h(\tau'-\rho)}{\sqrt{\tau'-\tau}}\,d\tau'$$

$$+ 2e^{(1-\varepsilon)\tau}\int_\tau^\infty e^{-(1-\varepsilon)\rho} f(\rho)\,d\rho \int_\tau^\rho \frac{h(\rho-\tau')}{\sqrt{\tau'-\tau}}\,d\tau'.$$

For $\rho < \tau$ we have

$$\int_\tau^\infty \frac{e^{-2\tau'} h(\tau'-\rho)}{\sqrt{\tau'-\tau}}\,d\tau' = 2e^{-2\tau}\int_0^\infty e^{-2x^2} h(x^2 + (\tau-\rho))$$

$$\leqslant 2e^{-2\tau}\int_0^\infty e^{-2x^2} h(x^2)\,dx < e^{-2\tau}\left(2 + a\sqrt{\frac{\pi}{2}}\right).$$

For $\rho > \tau$,

$$\int_\rho^\infty \frac{e^{-2\tau'} h(\tau'-\rho)}{\sqrt{\tau'-\tau}}\,d\tau' < 2e^{-2\rho}\int_0^\infty e^{-2x^2} h(x^2)\,dx < e^{-2\rho}\left(2 + a\sqrt{\frac{\pi}{2}}\right).$$

Finally, for $\rho > \tau$ we have

$$\int_\tau^\rho \frac{h(\rho-\tau')}{\sqrt{\tau'-\tau}}\,d\tau' = 2\sqrt{\rho-\tau}\int_0^1 h((\rho-\tau)(1-y^2))\,dy \leqslant 2a_1\sqrt{\rho-\tau}$$

$$+\begin{cases} -\sqrt{\rho-\tau}\ln(\rho-\tau) & \text{for } \rho-\tau\leqslant 1 \\ 0 & \text{for } \rho-\tau > 1. \end{cases}$$

Making use of the three latter inequalities, we obtain

$$\hat{X}\hat{Y}\hat{X}f(\tau) \leqslant 2\left(2 + a\sqrt{\frac{\pi}{2}}\right) e^{-(1+\varepsilon)\tau}\int_0^\tau e^{(1+\varepsilon)\rho} f(\rho)\,d\rho$$

$$+2\left(2 + a\sqrt{\frac{\pi}{2}}\right) e^{(1-\varepsilon)\tau}\int_\tau^\infty e^{-(1-\varepsilon)\rho} f(\rho)\,d\rho + 4a_1 e^{(1-\varepsilon)\tau}\int_\tau^\infty \sqrt{\rho-\tau}\, e^{-(1-\varepsilon)\rho} f(\rho)\,d\rho$$

$$+ 2e^{(1-\varepsilon)\tau}\int_\tau^{\tau+1} e^{-(1-\varepsilon)\rho} f(\rho)\,[-\sqrt{\rho-\tau}\ln(\rho-\tau)]\,d\rho$$

$$\leqslant 2\left(2 + a\sqrt{\frac{\pi}{2}}\right)\int_0^\infty f(\rho)\,d\rho + \frac{4a_1}{\sqrt{1-\varepsilon}}\int_\tau^\infty [\sqrt{(1-\varepsilon)(\rho-\tau)}\, e^{-(1-\varepsilon)(\rho-\tau)}]\, f(\rho)\,d\rho$$

$$+ 4\int_\tau^{\tau+1} e^{-(1-\varepsilon)(\rho-\tau)} f(\rho)\,[-\sqrt{\rho-\tau}\ln\sqrt{\rho-\tau}]\,d\rho.$$

Now let us use the fact that $\sqrt{xe^{-x}} \le 1/\sqrt{2e}$ for $x \ge 0$ and that $0 < -x\ln x \le 1/e$ for $0 < x < 1$. We obtain

$$\hat{X}\hat{Y}\hat{X}f(\tau) \leqslant 2\left(2 + a\sqrt{\frac{\pi}{2}}\right)\int_0^\infty f(\rho)\,d\rho + \frac{4a_1}{\sqrt{2e(1-\varepsilon)}}\int_\tau^\infty f(\rho)\,d\rho$$

$$+\frac{4}{e}\int_\tau^{\tau+1} f(\rho)\,d\rho. \tag{74}$$

Similar considerations yield the inequality

$$\hat{Y}\hat{X}\hat{Y}f(\tau) \leqslant 2\left(2 + a\sqrt{\frac{\pi}{2}}\right)\int_0^\infty f(\rho)\,d\rho + \frac{4a_1}{\sqrt{2e(1+\varepsilon)}}\int_0^\tau f(\rho)\,d\rho$$

$$+\frac{4}{e}\int_{\max\{0,\,\tau-1\}}^{\tau} f(\rho)\,d\rho. \tag{75}$$

The boundedness of $f_3(\tau)$ on $(0, \infty)$ now follows in self-evident fashion from estimates (68)–(75). The first statement of the theorem is proved.

To prove the second statement we note that

$$\hat{A}^4x(\tau, \omega) = \hat{A}x_3(\tau, \omega),\ \|x_3(\tau)\| \leqslant k_0e^{\varepsilon\tau},\ k_0 = \sup_{\tau>0} f_3(\tau) < \infty.$$

For $\omega\mathbf{n} > 0$,

$$|\hat{A}^4x(\tau, \omega)| \leqslant \frac{\sqrt{2\pi}\,[g]\,k_0}{\omega\mathbf{n}}\int_0^\tau e^{-\frac{\tau-\rho}{\omega\mathbf{n}}+\varepsilon\rho}\,d\rho \leqslant \frac{\sqrt{2\pi}\,[g]\,k_0}{1-|\varepsilon|}e^{\varepsilon\tau}.$$

For $\omega\mathbf{n} < 0$,

$$|\hat{A}^4x(\tau, \omega)| \leqslant \frac{\sqrt{2\pi}\,[g]\,k_0}{|\omega\mathbf{n}|}\int_\tau^\infty e^{-\frac{\rho-\tau}{|\omega\mathbf{n}|}+\varepsilon\rho}\,d\rho \leqslant \frac{\sqrt{2\pi}\,[g]\,k_0}{1-|\varepsilon|}e^{\varepsilon\tau}.$$

Finally, for $\omega\mathbf{n} = 0$,

$$|\hat{A}^4x(\tau, \omega)| = |x_\omega^*\,\hat{g}x_3(\tau)| \leqslant \sqrt{2\pi}\,[g]\,k_0e^{\varepsilon\tau} \leqslant \frac{\sqrt{2\pi}\,[g]\,k_0}{1-|\varepsilon|}e^{\varepsilon\tau}.$$

The theorem is proved.

Theorem 22. *Let g be an indicatrix and let $\hat{A} = \hat{A}_g$. The class $\mathfrak{B}_1$ is then not invariant with respect to the transformation $\hat{A}$.*

Proof. The indicatrix g satisfies Condition **I** (§3). By virtue of (14) this means that $g_0 > 0$. We denote by χ a function on Ω which is everywhere equal to unity, i.e. $\chi(\omega) = 1$, $\omega \in \Omega$. Let $f \in \mathfrak{U}$ (compare the proof of Theorem 21), and $x(\tau) = e^\tau f(\tau)\chi$. Clearly, $x(\cdot) \in \mathfrak{B}_1$. For $\omega \in \Omega$, $\tau > 0$ we have

$$x_\omega^*(\hat{g}x(\tau)) = e^\tau f(\tau)\,x_\omega^*(\hat{g}\chi) = g_0e^\tau f(\tau).$$

Let us set $z(\cdot) = \hat{A}(x(\cdot))$. Then

$$z(\tau, \omega) = \begin{cases} \dfrac{g_0}{|\omega n|}\displaystyle\int_\tau^\infty e^{-\frac{\rho-\tau}{|\omega n|}+\rho} f(\rho)\,d\rho & \text{for } \omega n < 0, \\ g_0 e^\tau f(\tau) & \text{for } \omega n = 0, \\ \dfrac{g_0}{\omega n}\displaystyle\int_0^\tau e^{-\frac{\tau-\rho}{\omega n}+\rho} f(\rho)\,d\rho & \text{for } \omega n > 0. \end{cases}$$

Making use of this formula and Lemma 1, we can readily compute the scalar product $(z(\tau), \chi)$,

$$(z(\tau), \chi) = 2\pi g_0 \int_0^\infty E_1(|\tau - \rho|)\, e^\rho f(\rho)\,d\rho,$$

where

$$E_1(k) = \int_1^\infty s^{-1} e^{-ks}\,ds, \quad k > 0.$$

This yields

$$\int_0^\infty e^{-\tau}(z(\tau), \chi)\,d\tau \geqslant 2\pi g_0 \int_0^\infty e^{-\tau}d\tau \int_\tau^\infty E_1(\rho - \tau)\, e^\rho f(\rho)\,d\rho$$
$$= 2\pi g_0 \int_0^\infty e^\rho f(\rho)\,d\rho \int_0^\rho e^{-\tau} E_1(\rho - \tau)\,d\tau.$$

But for $\rho > 0$ we have

$$e^\rho \int_0^\rho e^{-\tau} E_1(\rho - \tau)\,d\tau = \rho \int_0^\infty \frac{dt}{t(t+\rho)}(1 - e^{-t}) > \left(1 - \frac{1}{e}\right)\rho \int_1^\infty \frac{dt}{t(t+\rho)}$$
$$= \left(1 - \frac{1}{e}\right)\ln(1 + \rho)$$

and

$$\int_0^\infty e^{-\tau}(z(\tau), \chi)\,d\tau \geqslant 2\pi g_0 \left(1 - \frac{1}{e}\right)\int_0^\infty f(\rho)\ln(1 + \rho)\,d\rho.$$

We note now that $(z(\tau), \chi) \leq 2\sqrt{\pi}\,\|z(\tau)\|$, $\tau > 0$. Hence

$$\int_0^\infty \|z(\tau)\| e^{-\tau}\,d\tau \geqslant g_0 \sqrt{\pi}\left(1 - \frac{1}{e}\right)\int_0^\infty f(\rho)\ln(1 + \rho)\,d\rho.$$

It is clear that in order to prove the theorem we need merely choose $f \in \mathfrak{U}$ such that $\int_0^\infty f(\rho)\ln(1 + \rho)\,d\rho = \infty$. These conditions are satisfied, for example, by the function

$$f(\rho) = \begin{cases} 0 & \text{for } \rho \in (0, 1], \\ \dfrac{1}{(1+\rho)\ln^2(1+\rho)} & \text{for } \rho > 1. \end{cases}$$

Theorem 22 admits a self-evident generalization, since in proving it we made essential use only of the inclusion $g \in L_2(-1, 1)$ and of the inequality $g_0 \neq 0$.

§9. The method of successive approximations. The homogeneous Milne problem

Throughout the present section we shall assume that g is a fixed scattering indicatrix, and that $\hat{A}$ is the corresponding operator, $\hat{A} = \hat{A}_g$. Because g is nonnegative, the operator $\hat{A}$ is monotonic, i.e. if $x(\cdot) \in \mathcal{B}_1$ and $x(\tau, \omega) \geq 0$ at all points $(\tau, \omega) \in (0, \infty) \times \Omega$, then it is also the case that $\hat{A}x(\tau, \omega) \geq 0$, $(\tau, \omega) \in (0, \infty) \times \Omega$.

We shall show that to each solution of the characteristic equation (15) we can assign in a specific way a solution of the homogeneous Milne problem. Having constructed all such solutions, we shall investigate their properties. We shall also show that the class $\mathcal{B} = \bigcup_{\epsilon<1} \mathcal{B}_\epsilon$ does not contain any other solutions of the homogeneous Milne problem.

Let $\phi \in L_2(\Omega)$, let λ be a complex number, and let $\lambda_1 \equiv \mathrm{Re}\lambda < 1$. We set

$$x(\tau) = e^{\lambda\tau}\varphi, \quad \tau \in (0, \infty).$$

Clearly, $x(\cdot) \in \mathcal{B}_\epsilon$ for every $\epsilon \in (\lambda_1, 1)$, so that the function $\hat{A}x(\cdot)$ has been determined. If we assume that $\lambda \in Z_0$ (compare §4), we can reduce $\hat{A}x(\cdot)$ to the following simple form:

$$\hat{A}x(\tau, \omega) = \frac{e^{\lambda\tau}}{1+\lambda\omega n}\hat{g}\varphi(\omega) - \frac{\eta(\omega n)}{1+\lambda\omega n} e^{-\frac{\tau}{\omega n}} \hat{g}\varphi(\omega), \tag{76}$$

$$(\tau, \omega) \in (0, \infty) \times \Omega, \ \lambda \in Z_0, \ \mathrm{Re}\, \lambda < 1.$$

Here $\eta(s)$ is the Heaviside step function,

$$\eta(s) = \begin{cases} 0 & \text{for } s \leqslant 0, \\ 1 & \text{for } s > 0. \end{cases}$$

To derive (76) we note that

$$x^*_\omega(\hat{g}x(\tau)) = e^{\lambda\tau} x^*_\omega(\hat{g}\varphi) \equiv e^{\lambda\tau}\hat{g}\varphi(\omega), \quad (\tau, \omega) \in (0, \infty) \times \Omega,$$

and then apply formulas (57)–(59).

We assign to every $\lambda \in (-1, 1)$ a pair of functions $x_\lambda(\cdot)$ and $p_\lambda(\cdot)$ according to the formulas

$$x_\lambda(\tau) = e^{\lambda\tau}\Phi_\lambda, \quad \tau > 0, \tag{77}$$

$$p_\lambda(\tau, \omega) = e^{-\tau/\omega n}\eta(\omega n)\Phi_\lambda(\omega), \quad (\tau, \omega) \in (0, \infty) \times \Omega. \tag{78}$$

The function $\Phi_\lambda(\omega)$ and the function $M(\lambda)$ appearing below are defined in Theorem 6.

We set

$$\Phi_\lambda^+ = \sup_{\omega\in\Omega} \Phi_\lambda(\omega), \quad \Phi_\lambda^- = \inf_{\omega\in\Omega} \Phi_\lambda(\omega), \quad \lambda\in(-1,1).$$

Theorem 6 implies that

$$0<\Phi_\lambda^- \leqslant \Phi_\lambda^+ < \infty, \quad \lambda\in(-1,1).$$

It is clear that $x_\lambda(\tau,\omega)$ is continuous with respect to (τ,ω) on $(0,\infty)\times\Omega$ and $x_\lambda(\cdot)\in\mathcal{B}_\epsilon$ for all $\epsilon\in(\lambda,1)$, $\lambda\in(-1,1)$. For $x(\cdot)=x_\lambda(\cdot)$ formula (76) becomes

$$\hat{A}x_\lambda(\tau,\omega) = M(\lambda)(x_\lambda(\tau,\omega) - p_\lambda(\tau,\omega)), \quad (\tau,\omega)\in(0,\infty)\times\Omega, \quad \lambda\in(-1,1). \tag{79}$$

Since $x_\lambda(\tau)$ is strongly continuous with respect to τ on $(0,\infty)$, the function $\hat{A}x_\lambda(\tau,\omega)$ is continuous with respect to the pair of variables (τ,ω) on $(0,\infty)\times\Omega$ (compare the corollary of Theorem 17). Equation (79) now implies that $p_\lambda(\tau,\omega)$ is also continuous with respect to (τ,ω) on $(0,\infty)\times\Omega$ for every $\lambda\in(-1,1)$. (This can also be readily verified directly.) Thus $p_\lambda(\tau)$ is strongly continuous and therefore strongly measurable with respect to τ on $(0,\infty)$. Recalling the self-evident inequality

$$0\leqslant p_\lambda(\tau,\omega)\leqslant \Phi_\lambda^+ e^{-\tau}, \quad (\tau,\omega)\in(0,\infty)\times\Omega, \quad \lambda\in(-1,1), \tag{80}$$

we conclude that $p_\lambda(\cdot)\in\mathcal{B}_\epsilon$ for all $\lambda\in(-1,1)$ for $\epsilon\in(-1,1)$.

The invariance of the classes $\mathcal{B}_\epsilon$, $|\epsilon|<1$, established in Theorem 20 enables us to determine $\hat{A}^n x_\lambda(\cdot)$ and $\hat{A}^n p_\lambda(\cdot)$ for all $n\geq 1$ and $\lambda\in(-1,1)$. Making use of formula (79), we can show by induction over n that

$$\frac{1}{M^n(\lambda)}\hat{A}^n x_\lambda(\tau,\omega) = x_\lambda(\tau,\omega) - \sum_{\nu=0}^{n-1}\frac{1}{M^\nu(\lambda)}\hat{A}^\nu p_\lambda(\tau,\omega), \tag{81}$$

$$(\tau,\omega)\in(0,\infty)\times\Omega, \quad \lambda\in(-1,1), \quad n=1,2,\ldots$$

(Here we make use of the usual notation $\hat{A}^0 x(\cdot)\equiv x(\cdot)$, $x(\cdot)\in\mathcal{B}_1$.)

All of the subsequent constructions are based on the following lemma.

Lemma 22. *Let $\mathcal{F}$ be a bounded closed portion of the halfline $(0,\infty)$, and let $\mathbf{1}(\tau,\omega)=1$ for $(\tau,\omega)\in(0,\infty)\times\Omega$. Then $\mathbf{1}(\cdot)\in\mathcal{B}_\epsilon$ for $\epsilon\in(0,1)$, and uniformly with respect to all $(\tau,\omega)\in\mathcal{F}\times\Omega$ we have*

[i] $\lim_{n=\infty} g_0^{-n}\hat{A}^n\mathbf{1}(\tau,\omega)=0$,

[ii] $\sum_{\nu=0}^{\infty} g_0^{-\nu}\hat{A}^\nu p_0(\tau,\omega)=1$,

[iii] $\lim_{N=\infty}\hat{A}[\sum_{\nu=N+1}^{\infty} g_0^{-\nu}\hat{A}^\nu p_0(\cdot)](\tau,\omega)=0$.

Proof. The inclusion $\mathbf{1}(\cdot)\in\mathcal{B}_\epsilon$, $\epsilon\in(0,1)$, is self evident.

Let us begin by considering the conservative case $g_0=1$. We note that

$M(0) = 1$, $x_0(\tau, \omega) = \mathbf{1}(\tau, \omega) = 1$, $(\tau, \omega) \in (0, \infty) \times \Omega$. From formula (81) we obtain

$$\hat{A}^n \mathbf{1}(\tau, \omega) = 1 - \sum_{\nu=0}^{n-1} \hat{A}^\nu p_0(\tau, \omega), \quad (\tau, \omega) \in (0, \infty) \times \Omega, \quad n \geqslant 1, \quad g_0 = 1. \tag{82}$$

This means that $\{\hat{A}^n \mathbf{1}(\tau, \omega)\}$ is a nonincreasing sequence and that the series $\sum_{\nu=0}^{\infty} \hat{A}^\nu p_0(\tau, \omega)$ converges at every point $(\tau, \omega) \in (0, \infty) \times \Omega$. Moreover,

$$\hat{A}^n \mathbf{1}(\tau, \omega) \leqslant 1, \quad \sum_{\nu=0}^{n-1} \hat{A}^\nu p_0(\tau, \omega) \leqslant 1, \quad (\tau, \omega) \in (0, \infty) \times \Omega, \ n \geqslant 1.$$

We choose a $\lambda \in (-1, 0)$. It is clear that $x_\lambda(\tau, \omega) \leq \Phi_\lambda^+$. Because $\hat{A}$ is monotonic, we have

$$\frac{1}{M^n(\lambda)} \hat{A}^n x_\lambda(\tau, \omega) \leqslant \frac{\Phi_\lambda^+}{M^n(\lambda)} \hat{A}^n \mathbf{1}(\tau, \omega) \leqslant \frac{\Phi_\lambda^+}{M^n(\lambda)}.$$

Since $M(\lambda) > 1$ (Theorem 9), we have that $\lim_{n=\infty} (1/M^n(\lambda)) \hat{A}^n x_\lambda(\tau, \omega) = 0$ uniformly with respect to all $(\tau, \omega) \in (0, \infty) \times \Omega$. Making use of formula (81), we now find that for $\lambda \in (-1, 0)$,

$$x_\lambda(\tau, \omega) = \sum_{\nu=0}^{\infty} \frac{1}{M^\nu(\lambda)} \hat{A}^\nu p_\lambda(\tau, \omega),$$

where the series on the right converges uniformly with respect to $(\tau, \omega) \in (0, \infty) \times \Omega$. We can rewrite this expression as

$$x_\lambda(\tau, \omega) = \sum_{\nu=0}^{\infty} \frac{1}{M^\nu(\lambda)} \hat{A}^\nu p_0(\tau, \omega) + \sum_{\nu=0}^{\infty} \frac{1}{M^\nu(\lambda)} \hat{A}^\nu (p_\lambda(\cdot) - p_0(\cdot))(\tau, \omega). \tag{83}$$

This transformation is valid, since the series $\sum_{\nu=0}^{\infty} (1/M^\nu(\lambda)) \hat{A}^\nu p_0(\tau, \omega)$ converges for every $(\tau, \omega) \in (0, \infty) \times \Omega$.

Let $\sigma_\lambda = \sup_{\omega \in \Omega} |\Phi_\lambda(\omega) - 1|$. By the corollary of Theorem 8, $\lim_{\lambda=0} \sigma_\lambda = 0$. At every point $(\tau, \omega) \in (0, \infty) \times \Omega$ we have

$$|(p_\lambda(\cdot) - p_0(\cdot))(\tau, \omega)| = e^{-\tau/\omega \mathbf{n}} \eta(\omega \mathbf{n}) |\Phi_\lambda(\omega) - 1| \leqslant \sigma_\lambda p_0(\tau, \omega).$$

Hence, simultaneously for all $\lambda \in (-1, 0)$, $(\tau, \omega) \in (0, \infty) \times \Omega$, we have

$$\left| \sum_{\nu=0}^{\infty} \frac{1}{M^\nu(\lambda)} \hat{A}^\nu (p_\lambda(\cdot) - p_0(\cdot))(\tau, \omega) \right|$$

$$\leqslant \sum_{\nu=0}^{\infty} \hat{A}^\nu |p_\lambda(\cdot) - p_0(\cdot)|(\tau, \omega) \leqslant \sigma_\lambda \sum_{\nu=0}^{\infty} \hat{A}^\nu p_0(\tau, \omega) \leqslant \sigma_\lambda$$

and

$$\lim_{\lambda=0-0} \sum_{\nu=0}^{\infty} \frac{1}{M^\nu(\lambda)} A^\nu (p_\lambda(\cdot) - p_0(\cdot))(\tau, \omega) = 0$$

uniformly with respect to $(\tau, \omega) \in (0, \infty) \times \Omega$.

We choose our $(\tau, \omega) \in (0, \infty) \times \Omega$ and set $a_\nu = \hat{A}^\nu p_0(\tau, \omega)$, $\nu = 0, 1, 2, \cdots$. Let $\xi = 1/M(\lambda)$. Then $\xi \in (0, 1)$, and

$$\sum_{\nu=0}^{\infty} \frac{1}{M^\nu(\lambda)} \hat{A}^\nu p_0(\tau, \omega) = \sum_{\nu=0}^{\infty} a_\nu \xi^\nu.$$

We have already proved the convergence of the series $\sum_{\nu=0}^{\infty} a_\nu$. By the second theorem of Abel we have $\lim_{\xi=1-0} \sum_{\nu=0}^{\infty} a_\nu \xi^\nu = \sum_{\nu=0}^{\infty} a_\nu$, or, which is the same thing,

$$\lim_{\lambda=0-0} \sum_{\nu=0}^{\infty} \frac{1}{M^\nu(\lambda)} \hat{A}^\nu p_0(\tau, \omega) = \sum_{\nu=0}^{\infty} \hat{A}^\nu p_0(\tau, \omega), \quad (\tau, \omega) \in (0, \infty) \times \Omega.$$

Finally, it is clear that for every fixed (τ, ω) we have

$$\lim_{\lambda=0-0} x_\lambda(\tau, \omega) = x_0(\tau, \omega) = 1.$$

Collecting the resulting limit relations, we obtain (compare (83))

$$\sum_{\nu=0}^{\infty} \hat{A}^\nu p_0(\tau, \omega) = 1, \quad (\tau, \omega) \in (0, \infty) \times \Omega, \ g_0 = 1. \tag{84}$$

From this and from (82) we infer that

$$\lim_{n=\infty} \hat{A}^n \mathbf{1}(\tau, \omega) = 0, \quad (\tau, \omega) \in (0, \infty) \times \Omega, \quad g_0 = 1. \tag{85}$$

By Theorem 17[i], $\hat{A}^n \mathbf{1}(\tau)$ is a strongly continuous function of τ, $n \geq 0$. Hence $\hat{A}^n \mathbf{1}(\tau, \omega) = \hat{A}(\hat{A}^{n-1}\mathbf{1}(\cdot))(\tau, \omega)$ is a continuous function of the pair of variables (τ, ω) on $(0, \infty) \times \Omega$, $n \geq 1$ (compare the corollary of Theorem 17). Thus $\{\hat{A}^n \mathbf{1}(\tau, \omega) | n = 1, 2, \cdots\}$ is a monotonic sequence of nonnegative functions continuous with respect to the pair of variables (τ, ω); the sequence tends to zero. This already enables us to assert on the basis of the Dini theorem [38] that $\hat{A}^n \mathbf{1}(\tau, \omega) \to 0$ as $n \to \infty$ uniformly with respect to all $(\tau, \omega) \in \mathfrak{F} \times \Omega$, where $\mathfrak{F}$ is an arbitrary compact portion of the halfline $(0, \infty)$.

Thus statement [i] of the lemma is proved in the conservative case. In the general case we set $\tilde{g}(\mu) = g_0^{-1} g(\mu)$, $\mu \in [-1, 1]$. Clearly, $\tilde{g}$ is some new indicatrix, where $\tilde{g}_0 = 1$. Let $\check{A} = \hat{A}_{\tilde{g}}$. Then $\check{A} = g_0^{-1}\hat{A}$, which means that $\check{A}^n = g_0^{-n}\hat{A}^n$, $n = 1, 2, \cdots$. By what we have proved,

$$\lim_{n=\infty} \check{A}^n \mathbf{1}(\tau, \omega) = 0,$$

or, which is the same thing,

$$\lim_{n=\infty} g_0^{-n} \hat{A}\mathbf{1}(\tau, \omega) = 0$$

uniformly with respect to all $(\tau, \omega) \in \mathfrak{F} \times \Omega$. Statement [i] of the lemma is proved in the general case.

Let us set $\lambda = 0$ in (81):

$$g_0^{-n}\hat{A}^n\mathbf{1}(\tau, \omega) = 1 - \sum_{\nu=0}^{n-1} g_0^{-\nu}\hat{A}^\nu p_0(\tau, \omega), \quad (\tau, \omega) \in (0, \infty) \times \Omega, \quad n = 1, 2, \cdots.$$

Taking the limit in this expression as $n \to \infty$, we find that

$$\sum_{\nu=0}^{\infty} g_0^{-\nu}\hat{A}^\nu p_0(\tau, \omega) = 1,$$

where the series on the left converges uniformly on every set of the form $\mathfrak{F} \times \Omega$, where $\mathfrak{F}$ is a compact portion of the halfline $(0, \infty)$. Finally,

$$g_0^{-1}\hat{A}\left(\sum_{\nu=N+1}^{\infty} g_0^{-\nu}\hat{A}^\nu p_0(\cdot)\right)(\tau, \omega) \equiv g_0^{-1}\hat{A}\left[\mathbf{1}(\cdot) - \sum_{\nu=0}^{N} g_0^{-\nu}\hat{A}^\nu p_0(\cdot)\right](\tau, \omega)$$

$$= g_0^{-1}\hat{A}\mathbf{1}(\tau, \omega) - \sum_{\nu=1}^{N+1} g_0^{-\nu}\hat{A}^\nu p_0(\tau, \omega) = 1 - p_0(\tau, \omega) - \sum_{\nu=1}^{N+1} g_0^{-\nu}\hat{A}^\nu p_0(\tau, \omega)$$

$$= 1 - \sum_{\nu=0}^{N+1} g_0^{-\nu}\hat{A}^\nu p_0(\tau, \omega) \to 0$$

as $N \to \infty$ uniformly on $\mathfrak{F} \times \Omega$. The lemma is proved.

Lemma 23. *Let $\mathfrak{F}$ be an arbitrary bounded closed portion of the set $(0, \infty)$. Let $\psi \in L_2(\Omega)$, $\psi \neq \vartheta$, let λ and s be complex numbers, $\lambda \in Z_0$, $\lambda_1 \equiv \mathrm{Re}\lambda < 1$, $|s| \geq g_0$, and let*

$$s(1 + \lambda\omega\mathrm{n})\psi(\omega) = \hat{g}\psi(\omega), \quad \omega \in \Omega.$$

This means, among other things, that $\psi \in C(\Omega)$ and $\psi^+ \equiv \sup_{\omega\in\Omega}|\psi(\omega)| < \infty$. We set

$$x(\tau, \omega) = e^{\lambda\tau}\psi(\omega), \quad p(\tau, \omega) = \psi(\omega)p_0(\tau, \omega),$$
$$(\tau, \omega) \in (0, \infty) \times \Omega.$$

Then the following assertions are valid.

[i] $x(\cdot) \in \mathfrak{B}_\epsilon$, $\epsilon \in (\lambda_1, 1)$; $p(\cdot) \in \mathfrak{B}_{\epsilon'}$, $\epsilon' \in (-1, 1)$.

[ii] *For all $n \geq 1$ and $(\tau, \omega) \in (0, \infty) \times \Omega$ we have*

$$s^{-n}\hat{A}^n x(\tau, \omega) = x(\tau, \omega) - \sum_{\nu=0}^{n-1} s^{-\nu}\hat{A}^\nu p(\tau, \omega).$$

[iii] *The sequence $\{s^{-n}\hat{A}^n x(\tau, \omega)\}$ converges uniformly with respect to all $(\tau, \omega) \in \mathfrak{F} \times \Omega$:*

$$\lim_{n=\infty} s^{-n}\hat{A}^n x(\tau, \omega) = y(\tau, \omega).$$

Here

$$y(\tau, \omega) = x(\tau, \omega) - \sum_{\nu=0}^{\infty} s^{-\nu}\hat{A}^\nu p(\tau, \omega).$$

The series on the right converges absolutely and uniformly with respect to all $(\tau, \omega) \in \mathfrak{F} \times \Omega$, *and*

$$\left|\sum_{\nu=0}^{\infty} s^{-\nu}\hat{A}^{\nu} p(\tau, \omega)\right| \leqslant \psi^{+}, \quad (\tau, \omega) \in (0, \infty) \times \Omega.$$

[iv] $y(\tau, \omega)$ *is continuous with respect to* (τ, ω) *on* $(0, \infty) \times \Omega$.

[v] $y(\cdot) \in \mathfrak{B}_{\epsilon}$ *for* $\epsilon \in (\lambda_1, 1)$.

[vi] *If* $\lambda_1 > 0$, *then there exists a* $\tau_1 > 0$ *such that* $y(\tau) \neq \vartheta$ *for* $\tau > \tau_1$. *If* $\lambda_1 \leq 0$, *then* $y(\tau) \equiv \vartheta$ *for* $\tau \in (0, \infty)$.

[vii] $\hat{A} y(\cdot) = s y(\cdot)$.

Proof. The continuity of $\psi(\omega)$ is ensured by Theorem 1. The inclusion $x(\cdot) \subset \mathfrak{B}_{\epsilon}$, $\epsilon \in (\lambda_1, 1)$, was justified at the beginning of the present section, where we also showed that $p_0(\cdot) \in \mathfrak{B}_{\epsilon'}$ for $\epsilon' \in (-1, 1)$. Since $p(\tau, \omega)$ differs from $p_0(\tau, \omega)$ by a continuous bounded factor, it follows that $p(\cdot) \in \mathfrak{B}_{\epsilon'}$, $\epsilon' \in (-1, 1)$. From (76) we obtain the equation

$$\frac{1}{s}\hat{A}x(\tau, \omega) = x(\tau, \omega) - p(\tau, \omega).$$

We can now prove [ii] readily by induction over n.

We note that

$$|s^{-\nu}\hat{A}^{\nu} p(\tau, \omega)| \leqslant \psi^{+} g_0^{-\nu}\hat{A}^{\nu} p_0(\tau, \omega), \quad (\tau, \omega) \in (0, \infty) \times \Omega, \quad \nu = 0, 1, 2, \ldots.$$

This means (compare Lemma 22) that the series

$$\sum_{\nu=0}^{\infty} s^{-\nu}\hat{A}^{\nu} p(\tau, \omega) = \sigma(\tau, \omega),$$

converges uniformly with respect to $(\tau, \omega) \in \mathfrak{F} \times \Omega$, where

$$|\sigma(\tau, \omega)| \leqslant \psi^{+} \sum_{\nu=0}^{\infty} g_0^{-\nu}\hat{A}^{\nu} p_0(\tau, \omega) = \psi^{+}, \quad (\tau, \omega) \in (0, \infty) \times \Omega.$$

Formula [ii] now enables us to conclude that the sequence $\{s^{-n}\hat{A}^{n} x(\tau, \omega)\}$ also converges uniformly on $\mathfrak{F} \times \Omega$, and its limit $y(\tau, \omega)$ satisfies the relation

$$y(\tau, \omega) = x(\tau, \omega) - \sigma(\tau, \omega), \quad (\tau, \omega) \in (0, \infty) \times \Omega.$$

Thus statement [iii] is proved.

Theorem 17 enables us to establish readily the continuity of all the functions $s^{-n}\hat{A}^{n} x(\tau, \omega)$, $n = 1, 2, \cdots$ with respect to (τ, ω) on $(0, \infty) \times \Omega$. Statement [iv] now follows from the uniform convergence of $\{s^{-n}\hat{A}^{n} x(\tau, \omega)\}$ on $\mathfrak{F} \times \Omega$ and from the arbitrariness of $\mathfrak{F}$.

The self-evident relations

$$\|x(\tau)\| = e^{\lambda_1\tau}\|\psi\|, \quad \|\sigma(\tau)\| \leqslant 2\sqrt{\pi}\psi^+, \quad \tau > 0,$$

imply that

$$\|y(\tau)\| \geqslant \|x(\tau)\| - \|\sigma(\tau)\| \geqslant e^{\lambda_1\tau}\|\psi\| - \lambda\sqrt{\pi}\psi^+.$$

for $\tau > 0$. If $\lambda_1 > 0$, then $\|y(\tau)\| > 0$ at least for $\tau > \tau_1 = (1/\lambda_1)\ln(2\psi^+\sqrt{\pi}/\|\psi\|) \times$ $\|y(\tau)\| > 0$. On the other hand, if $\lambda_1 \leq 0$, then $|x(\tau, \omega)| \leq \psi^+$, $(\tau, \omega) \in (0, \infty) \times \Omega$. Hence

$$|s^{-n}\hat{A}^n x(\tau, \omega)| \leqslant \psi^+ g_0^{-n}\hat{A}^n \mathbf{1}(\tau, \omega), \quad (\tau, \omega) \in (0, \infty) \times \Omega,$$

and, by Lemma 22,

$$y(\tau, \omega) = \lim_{n=\infty} s^{-n}\hat{A}^n x(\tau, \omega) = 0, \quad (\tau, \omega) \in (0, \infty) \times \Omega.$$

Statement [vi] has been proved.

Let $\epsilon \in (\lambda_1, 1)$. If $\lambda_1 \leq 0$ then $y(\tau) \equiv \vartheta$, and clearly $y(\cdot) \in \mathcal{B}_\epsilon$. Let $\lambda_1 > 0$. We have

$$y(\cdot) = x(\cdot) - \sigma(\cdot),$$

where, by [i], $x(\cdot) \in \mathcal{B}_\epsilon$. By [iv], $\sigma(\tau, \omega)$ is a continuous function of $(\tau, \omega) \in (0, \infty) \times \Omega$. This means that $\sigma(\tau)$ is strongly continuous with respect to τ, and therefore strongly measurable. $\|\sigma(\tau)\|$ is bounded over the entire halfline $(0, \infty)$, and, since $\epsilon > 0$, we find that $\int_0^\infty e^{-\epsilon\tau}\|\sigma(\tau)\|\,d\tau < \infty$. Thus it is also the case that $\sigma(\cdot) \in \mathcal{B}_\epsilon$, which means that $y(\cdot) \in \mathcal{B}_\epsilon - \mathcal{B}_\epsilon \subset \mathcal{B}_\epsilon$.

It remains for us to prove [vii]. Let N be a natural number, and let $\rho_N(\tau, \omega) = \sum_{\nu=N+1}^\infty s^{-\nu}\hat{A}^\nu p(\tau, \omega)$. Then

$$y(\tau, \omega) = x(\tau, \omega) - \sum_{\nu=0}^{N} s^{-\nu}\hat{A}^\nu p(\tau, \omega) - \rho_N(\tau, \omega).$$

From this we infer that (compare [ii])

$$s^{-1}\hat{A}y(\tau, \omega) = s^{-1}\hat{A}x(\tau, \omega) - \sum_{\nu=1}^{N+1} s^{-\nu}\hat{A}^\nu p(\tau, \omega) - s^{-1}\hat{A}\rho_N(\tau, \omega)$$

$$= x(\tau, \omega) - \sum_{\nu=0}^{N+1} s^{-\nu}\hat{A}^\nu p(\tau, \omega) - s^{-1}\hat{A}\rho_N(\tau, \omega) = y(\tau, \omega) + r_N(\tau, \omega),$$

$$r_N(\tau, \omega) = \rho_{N+1}(\tau, \omega) - s^{-1}\hat{A}\rho_N(\tau, \omega).$$

Clearly,

$$|r_N(\tau, \omega)| \leqslant \psi^+ \sum_{\nu=N+2}^{\infty} g_0^{-\nu}\hat{A}^\nu p_0(\tau, \omega) + \frac{\psi^+}{|s|}\hat{A}\Big[\sum_{\nu=N+1}^{\infty} g_0^{-\nu}\hat{A}^\nu p_0(\cdot)\Big](\tau, \omega),$$

and, as we infer from Lemma 22,

$$\lim_{N=\infty} r_N(\tau, \omega) = 0, \quad (\tau, \omega) \in (0, \infty) \times \Omega.$$

This means that

$$s^{-1}\hat{A}y(\tau, \omega) = y(\tau, \omega), \quad (\tau, \omega) \in (0, \infty) \times \Omega.$$

The lemma is proved.

Lemma 24. *Let $\mathfrak{F}$ be an arbitrary bounded closed portion of the halfline $(0, \infty)$, and let*

$$x^{(0)}(\tau, \omega) = \tau$$

for $(\tau, \omega) \in (0, \infty) \times \Omega$. Then the following statements are valid.

[i] $x^{(0)}(\cdot) \in \mathfrak{B}_\epsilon$, $\epsilon \in (0, 1)$.

[ii] *The sequence $\{g_0^{-n}\hat{A}^n x^{(0)}(\tau, \omega)\}$ converges uniformly with respect to $(\tau, \omega) \in \mathfrak{F} \times \Omega$:*

$$\lim_{n=\infty} g_0^{-n}\hat{A}^n x^{(0)}(\tau, \omega) = y_0(\tau, \omega).$$

Here

$$y_0(\tau, \omega) = \tau + g_0\frac{1-\omega n}{g_0 - g_1} - \sum_{\nu=0}^{\infty} g_0^{-\nu}\hat{A}^\nu q(\tau, \omega),$$

$$(\tau, \omega) \in (0, \infty) \times \Omega,$$

where

$$q(\tau, \omega) = g_0(g_0 - g_1)^{-1}(1 - \omega n)\, p_0(\tau, \omega),$$

$q(\cdot) \in \mathfrak{B}_{\epsilon'}$, $\epsilon' \in (-1, 1)$, *and the series on the right converges uniformly with respect to $(\tau, \omega) \in \mathfrak{F} \times \Omega$. For all $(\tau, \omega) \in (0, \infty) \times \Omega$ we have*

$$0 \leqslant \sum_{\nu=0}^{\infty} g_0^{-\nu}\hat{A}^\nu q(\tau, \omega) \leqslant \frac{g_0}{g_0 - g_1}.$$

[iii] $y_0(\tau, \omega)$ *is continuous with respect to (τ, ω) and nonnegative on $(0, \infty) \times \Omega$.*

[iv] $y_0(\cdot) \in \mathfrak{B}_\epsilon$, $\epsilon \in (0, 1)$.

[v] $\hat{A}y_0(\cdot) = g_0 y_0(\cdot)$.

Proof. Inclusion [i] is self-evident. Let us compute $\hat{A}x^{(0)}(\tau, \omega)$. We have

$$x_\omega^*(\hat{g}x^{(0)}(\tau)) = g_0\tau, \quad (\tau, \omega) \in (0, \infty) \times \Omega.$$

Making use of fomulas (57)–(59), we obtain

$$\hat{A}x^{(0)}(\tau, \omega) = g_0(\tau - \omega n) + g_0\omega n p_0(\tau, \omega), \quad (\tau, \omega) \in (0, \infty) \times \Omega.$$

Now let $x^{(1)}(\tau, \omega) = \omega n$, $(\tau, \omega) \in (0, \infty) \times \Omega$. It is clear that $x^{(1)}(\cdot) \in \mathfrak{B}_\epsilon$, $\epsilon \in (0, 1)$,

$$x^*_\omega(\hat{g}x^{(1)}(\tau)) = x^*_\omega(g_1x^{(1)}(\tau)) = g_1\omega\mathbf{n}.$$

Hence

$$\hat{A}x^{(1)}(\tau,\ \omega) = g_1\,\omega\mathbf{n} - g_1\,\omega\mathbf{n}\,p_0(\tau,\ \omega),\quad (\tau,\ \omega) \in (0,\ \infty) \times \Omega.$$

We also note that

$$\hat{A}\mathbf{1}(\tau,\ \omega) = g_0 - g_0p_0(\tau,\ \omega) \quad \text{(compare (81)).}$$

Let us set $h = g_0(g_0 - g_1)^{-1}$, and let

$$x^{(2)}(\cdot) \doteq x^{(0)}(\cdot) + h\mathbf{1}(\cdot) - hx^{(1)}(\cdot).$$

Then $x^{(2)}(\cdot) \in \mathcal{B}_\epsilon$, $\epsilon \in (0,\ 1)$, and for all $(\tau,\ \omega) \in (0,\ \infty)$ we have

$$x^{(2)}(\tau,\ \omega) = \tau + h(1 - \omega\mathbf{n}),$$

$$g_0^{-1}\hat{A}x^{(2)}(\tau,\ \omega) = \tau + h(1 - \omega\mathbf{n}) - h(1 - \omega\mathbf{n})\,p_0(\tau,\ \omega) = x^{(2)}(\tau,\ \omega) - q(\tau,\ \omega).$$

It is clear that $q(\cdot) \in \mathcal{B}_{\epsilon'}$ for $\epsilon' \in (-1,\ 1)$. By induction over n we can now readily establish that

$$g_0^{-n}\hat{A}^nx^{(2)}(\tau,\ \omega) = x^{(2)}(\tau,\ \omega) - \sum_{\nu=0}^{n-1} g_0^{-\nu}\hat{A}^\nu q(\tau,\ \omega)$$

for $n \geq 1$, $(\tau,\ \omega) \in (0,\ \infty) \times \Omega$.

It is clear that for all $(\tau,\ \omega) \in (0,\ \infty) \times \Omega$ we have

$$0 \leqslant q(\tau,\ \omega) \leqslant hp_0(\tau,\ \omega).$$

Hence, for $\nu \geq 0$,

$$0 \leqslant g_0^{-\nu}\hat{A}^\nu q(\tau,\ \omega) \leqslant hg_0^{-\nu}\hat{A}^\nu p_0(\tau,\ \omega).$$

Lemma 22 now implies that the series

$$\sum_{\nu=0}^{\infty} g_0^{-\nu}\hat{A}^\nu q(\tau,\ \omega) = \sigma_0(\tau,\ \omega)$$

converges unifomly on $\mathcal{F} \times \Omega$, and that

$$0 \leqslant \sigma_0(\tau,\ \omega) \leqslant h$$

everywhere on $(0,\ \infty) \times \Omega$. We now see that

$$\lim_{n=\infty} g_0^{-n}\hat{A}^nx^{(2)}(\tau,\ \omega) \equiv y_0(\tau,\ \omega) = x^{(2)}(\tau,\ \omega) - \sigma_0(\tau,\ \omega)$$

uniformly with respect to $(\tau, \omega) \in \mathfrak{F} \times \Omega$.

Let $\delta(\cdot) = x^{(0)}(\cdot) - x^{(2)}(\cdot)$. Then $\delta(\cdot) \in \mathcal{B}_\epsilon$, $\epsilon \in (0, 1)$, and $|\delta(\tau, \omega)| \leq 2h$ everywhere on $(0, \infty) \times \Omega$. This means that

$$|g_0^{-n}\hat{A}^n\delta(\tau, \omega)| \leqslant 2hg_0^{-n}\hat{A}^n\mathbf{1}(\tau, \omega), \quad (\tau, \omega) \in (0, \infty) \times \Omega.$$

This in turn implies (compare Lemma 22) that

$$\lim_{n=\infty} g_0^{-n}\hat{A}^n\delta(\tau, \omega) = 0$$

uniformly on $\mathfrak{F} \times \Omega$. Since $g_0^{-n}\hat{A}^n x^{(0)}(\cdot) = g_0^{-n}\hat{A}^n x^{(2)}(\cdot) + g_0^{-n}\hat{A}^n\delta(\cdot)$, it follows that

$$\lim_{n=\infty} g_0^{-n}\hat{A}^n x^{(0)}(\tau, \omega) = y_0(\tau, \omega)$$

uniformly with respect to $(\tau, \omega) \in \mathfrak{F} \times \Omega$, and statement [ii] is proved.

Making use of Theorem 17, we can readily verify that all of the functions $\hat{A}^n x^{(0)}(\tau, \omega)$, $n = 0, 1, 2, \cdots$ are continuous with respect to (τ, ω) on $(0, \infty) \times \Omega$. Availing ourselves of the arbitrariness of $\mathfrak{F}$ and of the uniform convergence of the sequence $\{g_0^{-n}\hat{A}^n x^{(0)}(\tau, \omega)\}$ on $\mathfrak{F} \times \Omega$, we find that $y_0(\tau, \omega)$ is continuous with respect to (τ, ω) on $(0, \infty) \times \Omega$. The nonnegativeness of $y_0(\tau, \omega)$ follows from the nonnegativeness of $x^{(0)}(\tau, \omega)$ and from the monotonicity of the operator $\hat{A}$. Thus [iii] is proved.

We now recall that $y_0(\cdot) = x^{(2)}(\cdot) - \sigma_0(\cdot)$. Hence

$$|y_0(\tau, \omega)| \leqslant \tau + 3h, \ (\tau, \omega) \in (0, \infty) \times \Omega.$$

Combining this inequality with [iii], we obtain [iv].

To prove [v] we write out the expressions

$$\begin{aligned} g_0^{-1}\hat{A}y_0(\tau, \omega) &= g_0^{-1}\hat{A}x^{(2)}(\tau, \omega) - g_0^{-1}\hat{A}\sigma_0(\tau, \omega) \\ &= [x^{(2)}(\tau, \omega) - q(\tau, \omega)] - \Big[\sum_{\nu=1}^{N+1} g_0^{-\nu}\hat{A}^\nu q(\tau, \omega) \\ &\qquad + g_0^{-1}\hat{A}\Big(\sum_{\nu=N+1}^{\infty} g_0^{-\nu}\hat{A}^\nu q(\cdot)\Big)(\tau, \omega)\Big] \\ &= x^{(2)}(\tau, \omega) - \sum_{\nu=0}^{N+1} g_0^{-\nu}\hat{A}^\nu q(\tau, \omega) - g_0^{-1}\hat{A}\Big(\sum_{\nu=N+1}^{\infty} g_0^{-\nu}\hat{A}^\nu q(\cdot)\Big)(\tau, \omega). \qquad (*) \end{aligned}$$

But

$$0 \leqslant \hat{A}\Big(\sum_{\nu=N+1}^{\infty} g_0^{-\nu}\hat{A}^\nu q(\cdot)\Big)(\tau, \omega) \leqslant h\hat{A}\Big(\sum_{\nu=N+1}^{\infty} g_0^{-\nu}\hat{A}^\nu p_0(\cdot)\Big)(\tau, \omega).$$

By Lemma 22 this implies that

$$\lim_{N=\infty} \hat{A}\Big(\sum_{\nu=N+1}^{\infty} g_0^{-\nu}\hat{A}^{\nu}q(\cdot)\Big)(\tau, \omega) = 0, \quad (\tau, \omega)\in(0, \infty)\times\Omega.$$

Taking the limit as $N\to\infty$ in the right side of (*), we obtain the equation

$$g_0^{-1}\hat{A}y_0(\tau, \omega) = x^{(2)}(\tau, \omega) - \sigma_0(\tau, \omega) = y_0(\tau, \omega).$$

The lemma is proved.

Lemmas 23 and 24 enable us to find the solutions of the homogeneous Milne problem.

Theorem 23. *Let $\mathfrak{F}$ be an arbitrary bounded closed portion of the halfline* $(0, \infty)$, $\mathbf{k}\in\mathfrak{N}$, $\mathbf{k}>0$, *and let* $\{\psi_{kp}\,|p = 1, 2, \cdots, p_k\}$ *be the basis of the half-space* H_k *constructed in Lemma 15. We set*

$$x_{kp}(\tau, \omega) = e^{k\tau}\psi_{kp}(\omega), \quad p = 1, 2, \ldots, p_k, \quad k\in\mathfrak{N}, \quad k>0.$$

Then, for every pair of indices (k, p), $k\in\mathfrak{N}$, $k>0$, $1\le p\le p_k$, *the following statements are valid.*

[i] $x_{kp}(\cdot)\in\mathfrak{B}_\epsilon$, $\epsilon\in(k, 1)$.

[ii] *The limit*

$$\lim_{n=\infty}\hat{A}^n x_{kp}(\tau, \omega) = y_{kp}(\tau, \omega), \quad (\tau, \omega)\in(0, \infty)\times\Omega$$

exists. The sequence $\{\hat{A}^n x_{kp}(\tau, \omega)\}$ *converges uniformly with respect to all* $(\tau, \omega)\in\mathfrak{F}\times\Omega$, *and*

$$|y_{kp}(\tau, \omega) - x_{kp}(\tau, \omega)| \leqslant \psi_{kp}^{+} \equiv \sup_{\omega'\in\Omega}|\psi_{kp}(\omega')| < \infty$$

everywhere on $(0, \infty)\times\Omega$.

[iii] $y_{kp}(\cdot)\in\mathfrak{B}_\epsilon$ *for* $\epsilon\in(k, 1)$; *and* $y_{kp}(\tau, \omega)$ *is real and continuous with respect to* (τ, ω) *on* $(0, \infty)\times\Omega$.

[iv] *There exists a* $\tau_{kp}>0$ *such that*

$$y_{kp}(\tau) \neq \vartheta$$

for $\tau>\tau_{kp}$.

[v] $y_{kp}(\cdot)$ *is the solution of the homogeneous Milne problem,*

$$\hat{A}y_{kp}(\cdot) = y_{kp}(\cdot).$$

Proof. Let us choose $k\in\mathfrak{N}$, $k>0$, and p, $1\le p\le p_k$, and let $s = 1$, $\lambda = k$ and $\psi = \psi_{kp}$. The numbers s, λ and the function ψ satisfy the conditions of Lemma 23. It is now easy to see that the above theorem is a special case of Lemma 23 to within notation.

Let us denote by $\mathfrak{X}'$ the set of solutions of the equation $\hat{A}x(\cdot) = x(\cdot)$ constructed in Theorem 23:

$$\mathfrak{X}' = \{y_{kp}(\cdot) \mid k \in \mathfrak{N},\ k > 0,\ p = 1, 2, \ldots, p_k\}.$$

In the conservative case ($g_0 = 1$) the Milne problem has another solution in addition to solutions from the system $\mathfrak{X}'$. This is the function $y_0(\cdot)$ of Lemma 24. (In fact, relation [v] of Lemma 24 for $g_0 = 1$ becomes $\hat{A}y_0(\cdot) = y_0(\cdot)$.) Let us incorporate this function into $\mathfrak{X}'$. This yields the new system of solutions

$$\mathfrak{X} = \begin{cases} \mathfrak{X}' & \text{for} \quad g_0 < 1, \\ \mathfrak{X}' \cup \{y_0(\cdot)\} & \text{for} \quad g_0 = 1. \end{cases}$$

We note that all the functions of the system $\mathfrak{X}$ are real (compare Lemma 24 [iii], Theorem 23 [ii] and Lemma 15 [i]).

The following theorem contains some of the properties of the system $\mathfrak{X}$.

Theorem 24. [i] $\mathfrak{X} \subset \overline{\mathfrak{B}}$.

[ii] *The system* $\mathfrak{X}$ *is linearly independent.*

[iii] *Let*

$$J(\cdot) = \begin{cases} y_0(\cdot), & \text{if} \quad g_0 = 1, \\ y_{\lambda_0, 1}(\cdot), & \text{if} \quad g_0 < 1. \end{cases}$$

Then $J(\tau, \omega) \geq 0$ *everywhere on* $(0, \infty) \times \Omega$. *If* $y(\cdot)$ *is a linear combination with real coefficients of a finite number of elements of the system* $\mathfrak{X}$, *and if* $y(\tau, \omega)$ *is of fixed sign on* $(0, \infty) \times \Omega$, *then* $y(\cdot) = \alpha J(\cdot)$, $\alpha =$ const.

[iv] $J(\tau, \omega)$ *does not depend on the azimuth angle: for every* $\tau > 0$,

$$\hat{P}_n^{(0)} J(\tau) = J(\tau).$$

Proof. The inclusion $\mathfrak{X} \subset \overline{\mathfrak{B}}$ follows from the fact that $y_0(\cdot)$ and every $y_{kp}(\cdot)$ belongs to one of the classes $\mathfrak{B}_\epsilon$, $\epsilon < 1$.

Let $\{k_\nu \mid \nu = 1, 2, \cdots, t\} \subset \mathfrak{N}$, $0 < k_1 < k_2 < \cdots < k_t$, and let

$$\alpha y_0(\tau) + \sum_{\nu=1}^{t} \sum_{p=1}^{p_{k_\nu}} \alpha_{\nu p} y_{k_\nu p}(\tau) = \vartheta, \qquad \tau > 0.$$

Let us assume that the numbers $\alpha_{\nu p}$ contain nonzero numbers. Without limiting generality we can assume that $\sum_{p=1}^{p_{k_t}} |\alpha_{tp}| > 0$. Lemma 24 and Theorem 23 imply that

$$\sum_{p=1}^{p_{k_t}} \alpha_{tp} \psi_{k_t p}(\omega) + \sum_{\nu=1}^{t-1} \sum_{p=1}^{p_{k_\nu}} \alpha_{\nu p} e^{-(k_t - k_\nu)\tau} \psi_{k_\nu p}(\omega) + \alpha \tau e^{-k_t \tau} + \delta(\tau, \omega) e^{-k_t \tau} = 0,$$

$$(\tau, \omega) \in (0, \infty) \times \Omega,$$

where $\delta(\tau,\omega)$ is a function bounded on $(0,\infty)\times\Omega$. Taking the limit in the latter expression as $\tau\longrightarrow\infty$, we obtain

$$\sum_{p=1}^{p_{k_t}} \alpha_{tp}\psi_{k_t p}(\omega)=0, \quad \omega\in\Omega.$$

But the vectors $\psi_{k_{tp}}$, $p=1,2,\cdots,p_{k_t}$ are linearly independent, which implies that

$$\sum_{p=1}^{p_{k_t}} |\alpha_{tp}|=0.$$

The resulting contradiction means that $\alpha_{\nu p}=0$, $\nu=1,2,\cdots,t$, $p=1,2,\cdots$ $\cdots,p_{k_\nu}$. This in turn implies that $\alpha=0$. Statement [ii] proved.

Let us show that $J(\tau,\omega)\geq 0$. If $g_0=1$, then $J(\tau,\omega)=y_0(\tau,\omega)\geq 0$ (Lemma 24 [iii]). But if $g_0<1$, then

$$J(\tau,\omega)=\lim_{n=\infty}\hat{A}^n x_{\lambda_0,1}(\tau,\omega),\ (\tau,\omega)\in(0,\infty)\times\Omega,$$

and

$$x_{\lambda_0,1}(\tau,\omega)=e^{\lambda_0\tau}\psi_{\lambda_0,1}(\omega)=|(\omega\mathbf{n},\Phi^2_{\lambda_0})|^{-1/2}e^{\lambda_0\tau}\Phi_{\lambda_0}(\omega)>0$$

everywhere on $(0,\infty)\times\Omega$ (compare (40)). This and the monotonicity of $\hat{A}$ imply the nonnegativeness of $J(\tau,\omega)$.

Now let $y(\cdot)$ be a real combination of elements of $\mathfrak{X}$, and let it be of fixed sign. We can express it as

$$y(\cdot)=\alpha J(\cdot)+\sum_{\nu=1}^{t}\sum_{p=1}^{p_{k_\nu}}\alpha_{\nu p}y_{k_\nu p}(\cdot),$$

$$k_\nu\in\mathfrak{N},\ \nu=1,2,\ldots,t;\ 0\leqslant\lambda_0<k_1<\ldots<k_t.$$

Let us suppose that $\Sigma_{\nu=1}^{t}\Sigma_{p=1}^{p_{k_\nu}}|\alpha_{\nu p}|>0$. We can assume that $\Sigma_{p=1}^{p_{k_t}}|\alpha_{tp}|>0$. Then

$$y(\tau,\omega)=e^{k_t\tau}[\psi(\omega)+\delta_1(\tau,\omega)],\ (\tau,\omega)\in(0,\infty)\times\Omega,$$

where $\psi=\Sigma_{p=1}^{p_{k_t}}\alpha_{tp}\psi_{k_{tp}}$ and $\lim_{\tau=\infty}\delta_1(\tau,\omega)=0$ uniformly with respect to all $\omega\in\Omega$. The linear independence of the vectors $\psi_{k_{tp}}$ means that $\psi\neq\vartheta$. At the same time $\psi(\omega)$ is a real function, $\psi\in H_{k_t}$, $k_t\in\mathfrak{N}$, $k_t>\lambda_0$. This implies immediately that $\psi(\omega)$ changes sign on the sphere Ω (Theorem 12). Thus the function $y(\tau,\omega)$ for any sufficiently large τ must assume both positive and negative values on Ω. This contradicts the choice of $y(\cdot)$, so that $\alpha_{\nu p}=0$, $\nu=1,2,\cdots,t$, $p=1,2,\cdots$ $\cdots,p_{k_\nu}$. We conclude that $y(\cdot)=\alpha J(\cdot)$. Statement [iii] is proved.

By Lemma 24 we see that

$$\lim_{\nu=\infty} g_0^{-\nu} \hat{A}^{\nu} x^{(0)}(\tau, \omega) = y_0(\tau, \omega).$$

uniformly with respect to $\omega \in \Omega$ for every fixed $\tau > 0$. Hence

$$\lim_{\nu=\infty} g_0^{-\nu} \hat{A}^{\nu} x^{(0)}(\tau) = y_0(\tau), \quad \tau > 0$$

in the metric of $L_2(\Omega)$ as well. But by the definition of $x^{(0)}(\tau)$ we have $\hat{P}_{\mathbf{n}}^{(0)} x^{(0)}(\tau) = x^{(0)}(\tau)$, $\tau > 0$ (compare Lemma 3[vii]). Making use of Lemma 21, we find by induction over ν that

$$\hat{P}_{\mathbf{n}}^{(0)} \hat{A}^{\nu} x^{(0)}(\tau) = \hat{A}^{\nu} x^{(0)}(\tau), \quad \tau > 0, \quad \nu \geqslant 1.$$

Hence

$$\hat{P}_{\mathbf{n}}^{(0)} y_0(\tau) = \lim_{\nu=\infty} g_0^{-\nu} \hat{P}_{\mathbf{n}}^{(0)} \hat{A}^{\nu} x^{(0)}(\tau) = \lim_{\nu=\infty} g_0^{-\nu} \hat{A}^{\nu} x^{(0)}(\tau) = y_0(\tau).$$

Thus for $g_0 = 1$ we have

$$\hat{P}_{\mathbf{n}}^{(0)} J(\tau) = \hat{P}_{\mathbf{n}}^{(0)} y_0(\tau) = y_0(\tau) = J(\tau), \quad \tau > 0.$$

If $g_0 < 1$, then for a fixed $\tau > 0$ in the metric of $L_2(\Omega)$ we have

$$\lim_{\nu=\infty} \hat{A}^{\nu} x_{\lambda_0, 1}(\tau) = J(\tau).$$

For every $\tau > 0$ we have

$$\hat{P}_{\mathbf{n}}^{(0)} x_{\lambda_0, 1}(\tau) = |(\omega\mathbf{n}, \Phi_{\lambda_0}^2)|^{-1/2} e^{\lambda_0 \tau} \hat{P}_{\mathbf{n}}^{(0)} \Phi_{\lambda_0} = x_{\lambda_0, 1}(\tau)$$

(compare Theorem 15). Again applying Lemma 21, we find that

$$\hat{P}_{\mathbf{n}}^{(0)} \hat{A}^{\nu} x_{\lambda_0, 1}(\tau) = \hat{A}^{\nu} x_{\lambda_0, 1}(\tau), \quad \tau > 0, \quad \nu \geqslant 1,$$

which means that

$$\hat{P}_{\mathbf{n}}^{(0)} J(\tau) = J(\tau), \quad \tau > 0, \quad g_0 < 1.$$

The theorem is proved.

The nonnegative solution $J(\cdot)$ of the Milne problem isolated in Theorem 24 is of special interest owing to its clear physical meaning: it represents the phase density of the distribution of the radiation diffused through the halfspace $\tau > 0$. We shall now show that this solution is positive at every phase point.

Theorem 25. *Let* $J(\cdot)$ *be the function defined in Theorem 24. Then* $J(\tau, \omega) > 0$ *for all* $(\tau, \omega) \in (0, \infty) \times \Omega$.

Proof. Let h (like g) be an indicatrix. We define the function $k(\mu) \equiv (g * h)(\mu)$, $\mu \in [-1, 1]$ by the expression

$$k(\mu) = \sum_{n=0}^{\infty} \frac{2n+1}{4\pi} g_n h_n P_n(\mu).$$

The series on the right converges uniformly, since $\Sigma_{n=0}^{\infty} n|g_n|^2 + \Sigma_{n=0}^{\infty} n|h_n|^2 < \infty$ (compare Theorem 2).

The function $k(\mu)$ is also an indicatrix. In fact, $k(\mu)$ is continuous on $[-1, 1]$, and $k_0 = 2\pi \int_{-1}^{1} k(\mu)\, d\mu = g_0 h_0 \in (0, 1]$. The nonnegativeness of $k(\mu)$ follows from the relation

$$k(\omega_1\omega_2) = \int_{\Omega} g(\omega_1\omega) h(\omega\omega_2)\, d\omega, \quad \omega_{1,2} \in \Omega,$$

which can be obtained as follows. Let $u(\omega) = h(\omega\omega_2)$, $\omega \in \Omega$, and let σ be an orthonormal basis of the space R_3. Then (compare Theorem 2)

$$\int_{\Omega} g(\omega_1\omega) h(\omega\omega_2)\, d\omega = \hat{g}u(\omega_1) = \sum_{n=0}^{\infty} g_n \sum_{m=-n}^{n} (u, Y_{\sigma n}^{(m)}) Y_{\sigma n}^{(m)}(\omega_1).$$

But

$$(u, Y_{\sigma n}^{(m)}) = \int_{\Omega} h(\omega_2\omega) \overline{Y_{\sigma n}^{(m)}(\omega)}\, d\omega = \overline{\hat{h}Y_{\sigma n}^{(m)}(\omega_2)} = h_n \overline{Y_{\sigma n}^{(m)}(\omega_2)}.$$

Hence (compare Lemma 2[iii])

$$\int_{\Omega} g(\omega_1\omega) h(\omega\omega_2)\, d\omega = \sum_{n=0}^{\infty} g_n h_n \sum_{m=-n}^{n} Y_{\sigma n}^{(m)}(\omega_1) \overline{Y_{\sigma n}^{(m)}(\omega_2)}$$

$$= \sum_{n=0}^{\infty} \frac{2n+1}{4\pi} g_n h_n P_n(\omega_1\omega_2) = k(\omega_1\omega_2).$$

It is easy to show that $\hat{k} = \hat{h}\hat{g} = \hat{g}\hat{h}$ (compare the proof of Lemma 4).

Just as we associated the operator $\hat{A} = \hat{A}_{\hat{g}}$ with the indicatrix g (cf. the Definition in §7), we associate the operators $\hat{A}_h$ and $\hat{A}_k$ with the indicatrixes h and k.

a) Let $x(\cdot) \in \overline{\mathfrak{B}}$, where $x(\tau, \omega)$ is nonnegative and continuous with respect to (τ, ω) on $(0, \infty) \times \Omega$. The conditions

$$(\tau_0, \omega_0) \in (0, \infty) \times \Omega, \quad \omega_0 \mathbf{n} \neq 0, \quad \hat{A}_k x(\tau_0, \omega_0) > 0$$

now imply that

$$\hat{A}_g \hat{A}_h\, x(\tau_0, \omega_0) > 0.$$

To prove this statement we set $y(\rho) = \hat{h}x(\rho)$, so that $y(\rho, \omega) = x_\omega^*(\hat{h}x(\rho))$, $(\rho, \omega) \in (0, \infty) \times \Omega$. Clearly, $y(\rho, \omega)$ is continuous on $(0, \infty) \times \Omega$. First, let us assume that $\omega_0\mathbf{n} > 0$. It is easy to see that $\hat{A}_g\hat{A}_h x(\tau_0, \omega_0)$ is positive if and only if the integral

$$\int_0^{\tau_0} d\tau \int_\Omega g(\omega_0\omega)\, \hat{A}_h x(\tau, \omega)\, d\omega \tag{86}$$

is positive.

Let us replace $\hat{A}_h x(\tau, \omega)$ in the integrand by a nonnegative function of simple form which vanishes wherever $\hat{A}_h x(\tau, \omega) = 0$. As such a function we can take, for example, one which equals

$$\begin{array}{ll} \int_0^{\tau} y(\rho, \omega)\, d\rho & \text{for} \quad \tau \in (0, \tau_0),\ \omega\mathbf{n} > 0; \\ 0 & \text{for} \quad \tau \in (0, \tau_0),\ \omega\mathbf{n} = 0; \\ \int_\tau^{\tau_0} y(\rho, \omega)\, d\rho & \text{for} \quad \tau \in (0, \tau_0),\ \omega\mathbf{n} < 0. \end{array}$$

This yields the expression

$$\int_0^{\tau_0} d\tau \int_{\omega\mathbf{n}>0} g(\omega_0\omega)\, d\omega \int_0^{\tau} y(\rho, \omega)\, d\rho + \int_0^{\tau_0} d\tau \int_{\omega\mathbf{n}<0} g(\omega_0\omega)\, d\omega \int_\tau^{\tau_0} y(\rho, \omega)\, d\rho. \tag{87}$$

(Making use of the fact that $y(\rho, \omega)$ is continuous, we can readily show that all of the integrals appearing in this expression have meaning.) If (87) is positive, then (86) is also positive, which means that $\hat{A}_g\hat{A}_h x(\tau_0, \omega_0) > 0$.

By some obvious transformations we can reduce (87) to

$$\int_0^{\tau_0} (\tau_0 - \rho)\, d\rho \int_{\omega\mathbf{n}>0} g(\omega_0\omega)\, y(\rho, \omega)\, d\omega + \int_0^{\tau_0} \rho\, d\rho \int_{\omega\mathbf{n}<0} g(\omega_0\omega)\, y(\rho, \omega)\, d\omega.$$

(The change in the order of integration is readily justifiable by means of the Fubini theorem and Lemma 1.) But the above sum is positive if and only if the quantity

$$\int_0^{\tau_0} d\rho \int_{\omega\mathbf{n}>0} g(\omega_0\omega)\, y(\rho, \omega)\, d\omega + \int_0^{\tau_0} d\rho \int_{\omega\mathbf{n}<0} g(\omega_0\omega)\, y(\rho, \omega)\, d\omega$$

$$= \int_0^{\tau_0} x^*_{\omega_0} (\hat{g} y(\rho))\, d\rho = \int_0^{\tau_0} x^*_{\omega_0} (\hat{k} x(\rho))\, d\rho$$

is positive.

Moreover, $\int_0^{\tau_0} x^*_{\omega_0} (\hat{k} x(\rho))\, d\rho > 0$ if and only if $\hat{A}_k x(\tau_0, \omega_0) > 0$. It is now clear that the conditions $\omega_0\mathbf{n} > 0$ and $\hat{A}_k x(\tau_0, \omega_0) > 0$ imply that $\hat{A}_g\hat{A}_h x(\tau_0, \omega_0) > 0$.

Let us assume now that $\omega_0\mathbf{n} < 0$, and $\hat{A}_k x(\tau_0, \omega_0) > 0$. Then, clearly,

$$0<\int_{\tau_0}^{\infty} e^{-\rho}\, x^{*}_{\omega_0}(\hat{k}x(\rho))\, d\rho=\int_{\tau_0}^{\infty} e^{-\rho}\, d\rho \int_{\omega\mathbf{n}>0} g(\omega_0\omega)\, y(\rho,\ \omega)\, d\omega$$

$$+\int_{\tau_0}^{\infty} e^{-\rho}\, d\rho \int_{\omega\mathbf{n}<0} g(\omega_0\omega)\, y(\rho,\ \omega)\, d\omega.$$

Hence

$$0<\int_{\tau_0}^{\infty} e^{-2\rho}\, d\rho \int_{\omega\mathbf{n}>0} g(\omega_0\omega)\, y(\rho,\ \omega)\, d\omega+\int_{\tau_0}^{\infty} e^{-\rho}(e^{-\tau_0}-e^{-\rho})\, d\rho \int_{\omega\mathbf{n}<0} g(\omega_0\omega)$$

$$\times\, y(\rho,\ \omega)\, d\omega=\int_{\tau_0}^{\infty} e^{-\tau}\, d\tau \int_{\omega\mathbf{n}>0} g(\omega_0\omega)\, d\omega \int_{\tau_0}^{\tau} e^{-\rho} y(\rho,\ \omega)\, d\rho$$

$$+\int_{\tau_0}^{\infty} e^{-\tau}\, d\tau \int_{\omega\mathbf{n}<0} g(\omega_0\omega)\, d\omega \int_{\tau}^{\infty} e^{-\rho} y(\rho,\ \omega)\, d\rho.$$

For this reason

$$\int_{\tau_0}^{\infty} e^{-\tau}\, d\tau \int_{\Omega} g(\omega_0\omega)\, \hat{A}_h x(\tau,\ \omega)\, d\omega>0,$$

and $\hat{A}_g \hat{A}_h x(\tau_0, \omega_0) > 0$. Assertion a) is proved.

b) Let $x(\cdot) \in \overline{\mathfrak{B}}$, $x'(\cdot) \in \overline{\mathfrak{B}}$, $x(\tau, \omega)$ and $x'(\tau, \omega)$ be nonnegative on $(0, \infty) \times \Omega$, and let $x(\tau, \omega) > 0$ for $\omega\mathbf{n} \neq 0$ imply that $x'(\tau, \omega) > 0$. Then, if $\hat{A}x(\tau, \omega) > 0$ at some point (τ, ω), then it is also the case that $\hat{A}x'(\tau, \omega) > 0$.

To prove b) we set $y(\tau, \omega) = x^*_\omega(\hat{g}x(\tau))$, $y'(\tau, \omega) = x^*_\omega(\hat{g}x'(\tau))$, $(\tau, \omega) \in (0, \infty) \times \Omega$. We obtain

$$y(\tau,\ \omega)=\int_{\Omega,\ \omega'\mathbf{n}\neq 0} g(\omega\omega')\, x(\tau,\ \omega')\, d\omega',$$

$$y'(\tau,\ \omega)=\int_{\Omega,\ \omega'\mathbf{n}\neq 0} g(\omega\omega')\, x'(\tau,\ \omega')\, d\omega',$$

which implies that $y'(\tau, \omega) > 0$ wherever $y(\tau, \omega) > 0$.

We assume now that $\hat{A}x(\tau_0, \omega_0) > 0$ at some point $(\tau_0, \omega_0) \in (0, \infty) \times \Omega$. If $\omega_0\mathbf{n} > 0$, then $\int_0^{\tau_0} y(\tau, \omega_0)d\tau > 0$, where, by what was said above, $\int_0^{\tau_0} y'(\tau, \omega_0)d\tau > 0$, which means that $\hat{A}x'(\tau_0, \omega_0) > 0$. If $\omega_0\mathbf{n} < 0$, then $\int_{\tau_0}^{\infty} e^{-\tau} y(\tau, \omega_0)d\tau > 0$, so that $\int_{\tau_0}^{\infty} e^{-\tau} y'(\tau, \omega_0)d\tau > 0$, and $\hat{A}x'(\tau_0, \omega_0) > 0$. Finally, if $\omega_0\mathbf{n} = 0$, then $y(\tau_0, \omega_0) = \hat{A}x(\tau_0, \omega_0) > 0$, which implies that $\hat{A}x'(\tau_0, \omega_0) = y'(\tau_0, \omega_0) > 0$. Assertion b) is proved.

Now let us set $g_{(1)} = g$ and define the indicatrix $g_{(p)}$ for every integer $p \geq 2$ as we did in Lemma 4. Clearly, $g_{(p+1)} = g * g_{(p)}$, $p \geq 1$.

c) Let $x(\cdot) \in \overline{\mathfrak{B}}$, and let $x(\tau, \omega)$ be nonnegative and continuous with respect to (τ, ω) on $(0, \infty) \times \Omega$; $p \geq 1$. If $(\tau_0, \omega_0) \in (0, \infty) \times \Omega$, $\omega_0 \mathbf{n} \neq 0$, and $\hat{A}_{g(p)} x(\tau_0, \omega_0) > 0$, then it is also the case that $\hat{A}^p x(\tau_0, \omega_0) > 0$.

Since $\hat{A}_{g(1)} = \hat{A}$, this statement is certainly true for $p = 1$. Let us assume that it is also true for some $p \geq 1$ and prove it for $p+1$. Let $(\tau_0, \omega_0) \in (0, \infty) \times \Omega$, $\omega_0 \mathbf{n} \neq 0$, and $\hat{A}_{g(p+1)} x(\tau_0, \omega_0) > 0$. Since $g_{(p+1)} = g * g_{(p)}$, we establish with the aid of a) that

$$\hat{A}(\hat{A}_{g(p)} x(\cdot))(\tau_0, \omega_0) > 0.$$

Now let us compare the functions $\hat{A}_{g(p)} x(\cdot)$ and $\hat{A}^p x(\cdot)$. It is clear that they both occur in the class $\overline{\mathfrak{B}}$. By the inductive assumption, at every point $(\tau, \omega) \in (0, \infty) \times \Omega$, $\omega \mathbf{n} \neq 0$, where $\hat{A}_{g(p)} x(\tau, \omega) > 0$, we have the inequality $\hat{A}^p x(\tau, \omega) > 0$. Making use of b), we infer from this that

$$\hat{A}(\hat{A}^p x(\cdot))(\tau_0, \omega_0) > 0,$$

or, equivalently, that $\hat{A}^{p+1} x(\tau_0, \omega_0) > 0$. Statement c) is proved.

It is now easy for us to show that $J(\tau, \omega)$ is positive. We choose a p such that $g_{(p)}(\mu) \geq \alpha > 0$, $\mu \in [-1, 1]$ (compare Theorem 4). From the definition of $J(\tau, \omega)$ we find that $\lim_{\tau=\infty} J(\tau, \omega) = \infty$ uniformly with respect to all $\omega \in \Omega$ (compare Theorem 24, Lemma 24 [ii] and Theorem 23 [ii]). We choose $\tau_1 > 0$ in such a way that when $\tau > \tau_1$ the function $J(\tau, \omega) > 1$ immediately for all $\omega \in \Omega$. Let $\tau > 0$, $\omega \in \Omega$, $\omega \mathbf{n} < 0$. Then

$$J(\tau, \omega) = \hat{A} J(\tau, \omega) = \frac{1}{|\omega \mathbf{n}|} \int_\tau^\infty e^{-\frac{\rho - \tau}{|\omega \mathbf{n}|}} y(\rho, \omega)\, d\rho,$$

where

$$y(\rho, \omega) = x^*_\omega(\hat{g} J(\rho)) = \int_\Omega g(\omega\omega') J(\rho, \omega')\, d\omega'.$$

If $\rho > \tau_1$, then

$$y(\rho, \omega) \geqslant \int_\Omega g(\omega\omega')\, d\omega' = g_0 > 0, \quad \omega \in \Omega,$$

which means that $J(\tau, \omega) > 0$ at every point $(\tau, \omega) \in (0, \infty) \times \Omega$, $\omega \mathbf{n} < 0$. Hence for $(\tau, \omega) \in (0, \infty) \times \Omega$ we have

$$x^*_\omega(\hat{g}_{(p)} J(\tau)) = \int_\Omega g_{(p)}(\omega\omega') J(\tau, \omega')\, d\omega' \geqslant \alpha \int_\Omega J(\tau, \omega')\, d\omega' > 0,$$

which means (cf. (57)–(59)) that $\hat{A}_{g(p)} J(\tau, \omega) > 0$ at every point $(\tau, \omega) \in (0, \infty) \times \Omega$. We now note that $J(\cdot)$ satisfies all the conditions of statement c) (by Lemma 24 [iii], Theorem 23 [iii] and Theorem 24 [iii]). Thus,

$$J(\tau, \omega) = \hat{A}^p J(\tau, \omega) > 0, \quad (\tau, \omega) \in (0, \infty) \times \Omega, \ \omega \mathbf{n} \neq 0.$$

It remains for us to prove that $J(\tau, \omega)$ is positive for $\omega \mathbf{n} = 0$.

Let $(\tau, \omega) \in (0, \infty) \times \Omega$.

$$\overset{*}{x}_\omega (\hat{g}J(\tau)) = \int_\Omega g(\omega\omega')\, J(\tau, \omega')\, d\omega' > 0,$$

since $J(\tau, \omega') > 0$ almost everywhere on Ω, and since $\int_\Omega g(\omega\omega')\,d\omega' > 0$. This means that

$$J(\tau, \omega) = \hat{A}J(\tau, \omega) = \overset{*}{x}_\omega (\hat{g}J(\tau)) > 0,$$

for $\omega\mathbf{n} = 0$, and the theorem is proved.

The system $\mathfrak{X}$ of solutions of the homogeneous Milne problem which we constructed in this section consists of functions which increase without limit as $\tau \to \infty$. It is easy to show that bounded and sufficiently slowly increasing functions cannot be solutions of the Milne problem. To formulate this result rigorously we introduce the following definition.

Definition. Let $\epsilon \in (-1, 1)$. We denote by $\mathfrak{T}_\epsilon$ the class of all functions $y(\cdot) \in \mathfrak{K}$, each of which can be represented as a sum

$$y(\cdot) = \sum_{\nu=1}^{3} y_\nu(\cdot),$$

where

$$y_1(\cdot) \in \mathfrak{B}_0,\ y_2(\cdot) \in \bigcup_{\varepsilon' < \varepsilon} \mathfrak{B}_{\varepsilon'},\ y_3(\cdot) \in \mathfrak{B}_1 \text{ and } \sup_{\tau > 0} \| y_3(\tau) \| < \infty.$$

It is clear that $\mathfrak{T}_\epsilon$ is a linear manifold in $\mathfrak{K}$, $\epsilon \in (-1, 1)$.

Lemma 25. *The class $\mathfrak{T}_{\lambda_0}$ does not contain nontrivial solutions of the homogeneous Milne problem.*

Proof. Let $y(\cdot) \in \mathfrak{T}_{\lambda_0}$, $\hat{A}y(\cdot) = y(\cdot)$, $y(\cdot) = y_1(\cdot) + y_2(\cdot) + y_3(\cdot)$, where $y_1(\cdot) \in \mathfrak{B}_0$, $y_2(\cdot) \in \bigcup_{\epsilon < \lambda_0} \mathfrak{B}_\epsilon$ and $\sup_{\tau > 0} \|y_3(\tau)\| = k_1 < \infty$. To prove the lemma we need merely show that $y(\tau, \omega) \equiv 0$ on $(0, \infty) \times \Omega$.

Theorem 17[v] implies that

$$|\hat{A}y_3(\tau, \omega)| \leqslant \sqrt{2\pi}\,[g]\,k_1,\quad (\tau, \omega) \in (0, \infty) \times \Omega,$$

so that

$$|\hat{A}^n y_3(\tau, \omega)| \leqslant \sqrt{2\pi}\,[g]\,k_1 \hat{A}^{n-1}\mathbf{1}(\tau, \omega),\quad n \geqslant 1.$$

Hence

$$\lim_{n=\infty} \hat{A}^n y_3(\tau, \omega) = 0,\quad (\tau, \omega) \in (0, \infty) \times \Omega$$

(compare Lemma 22).

The function $y_1(\cdot)$ satisfies the following inequality, which is readily obtainable from Theorem 21:

$$|\hat{A}^n y_1(\tau, \omega)| \leqslant k_2 \hat{A}^{n-4} \mathbf{1}(\tau, \omega), \ n \geqslant 1, \ 0 < k_2 = \text{const} < \infty.$$

This means that $\lim_{n=\infty} \hat{A}^n y_1(\tau, \omega) = 0$, $(\tau, \omega) \in (0, \infty) \times \Omega$.

Finally, if $g_0 = 1$, then $y_2(\cdot) \in \mathfrak{B}_0$, and, repeating the argument for the function $y_1(\cdot)$, we obtain

$$\lim_{n=\infty} \hat{A}^n y_2(\tau, \omega) = 0, \ (\tau, \omega) \in (0, \infty) \times \Omega.$$

On the other hand, if $g_0 < 1$, then $y_2(\cdot) \in \mathfrak{B}_\epsilon$ for some ϵ, $0 \leq \epsilon < \lambda_0$, and (compare Theorem 21) we have

$$|\hat{A}^4 y_2(\tau, \omega)| \leqslant k_3 e^{\varepsilon\tau} \leqslant \frac{k_3}{\Phi_\varepsilon^-} x_\varepsilon(\tau, \omega), (\tau, \omega) \in (0, \infty) \times \Omega, \ 0 < k_3 = \text{const} < \infty.$$

This and formula (81) for $n \geq 4$ yield

$$|\hat{A}^n y_2(\tau, \omega)| \leqslant \frac{k_3}{\Phi_\varepsilon^-} \hat{A}^{n-4} x_\varepsilon(\tau, \omega) \leqslant M^{n-4}(\varepsilon) \frac{k_3}{\Phi_\varepsilon^-} x_\varepsilon(\tau, \omega).$$

But $0 < M(\epsilon) < M(\lambda_0) = 1$ (compare Theorem 9), so that

$$\lim_{n=\infty} |\hat{A}^n y_2(\tau, \omega)| = 0, \ (\tau, \omega) \in (0, \infty) \times \Omega.$$

Applying the limit relations derived above and the self-evident inequality

$$|y(\tau, \omega)| = |\hat{A}^n y(\tau, \omega)| \leqslant \sum_{\nu=1}^{3} |\hat{A}^n y_\nu(\tau, \omega)|, \ n \geqslant 1,$$

we obtain

$$y(\tau, \omega) = 0, \ (\tau, \omega) \in (0, \infty) \times \Omega.$$

In §11 we shall show that the class $\overline{\mathfrak{B}}$ does not contain any solutions of the homogeneous Milne problem other than the functions from the system $\mathfrak{X}$ and their linear combinations. This enables us to strengthen Lemma 25 in the obvious way.

§10. The inhomogeneous Milne problem. Problem I

In this section we consider those solutions of inhomogeneous equation (52) which can be represented by means of the appropriate Neumann series. We already know that the homogeneous Milne problem always has a nontrivial solution, which means that the solution of inhomogeneous equation (52), if it exists at all, is definitely nonunique. However, it is possible in a certain natural way to isolate a class of functions which do not increase too rapidly at infinity and within whose limits equation (52) admits of exactly one solution (Theorem 26).

Problems of the (52) type without radiation sources in the halfspace $\tau > 0$ under inhomogeneous boundary conditions are of special interest. The absolute term of equation (52) in such problems becomes (cf. §1)

$$f(\tau, \omega) = B(\omega) e^{-\tau/\omega n}, \quad (\tau, \omega) \in (0, \infty) \times \Omega,$$

where $B(\omega)$ is the directional distribution of the radiation falling from the outside onto the boundary $\tau = 0$; $B(\omega) = 0$ for $\omega n \leq 0$. The solutions of problems of this type have certain special features which we shall consider in this section.

Lemma 26. *Let $g_0 < 1$, $\epsilon \in (-\lambda_0, \lambda_0)$, and let $\mathfrak{F}$ be a bounded closed portion of the halfline $(0, \infty)$, $f(\cdot) \in \mathfrak{B}_\epsilon$. Then*

[i] *The series*

$$\sum_{n=0}^{\infty} \hat{A}^n f(\tau, \omega) = s(\tau, \omega)$$

converges uniformly with respect to all $(\tau, \omega) \in \mathfrak{F} \times \Omega$ at each point $(\tau, \omega) \in (0, \infty) \times \Omega$. Being an abstract function of the variable $\tau \in (0, \infty)$, the sum $s(\cdot)$ of this series has the following properties:

[ii] *$s(\cdot) \in \mathfrak{T}_{\lambda_0}$, and $s(\cdot) \in \mathfrak{B}_{\epsilon'}$ for every $\epsilon' \in (\epsilon, 1)$.*

[iii] *$s(\cdot) = x(\cdot) + u(\cdot)$, where $x(\cdot) = \sum_{n=0}^{3} \hat{A}^n f(\cdot) \in \mathfrak{B}_\epsilon$, and the numerical function $u(\tau, \omega)$ is continuous with respect to the pair of variables (τ, ω) on $(0, \infty) \times \Omega$, where*

$$\sup_{(\tau, \omega) \in (0, \infty) \times \Omega} e^{-\varepsilon\tau} |u(\tau, \omega)| < \infty.$$

[iv] *$s(\cdot) = \hat{A}s(\cdot) + f(\cdot)$.*

Proof. For $n \geq 4$,

$$\hat{A}^n f(\cdot) = \hat{A}^{n-4} f_1(\cdot), \quad f_1(\cdot) = \hat{A}^4 f(\cdot).$$

There exists a positive number $k < \infty$ such that

$$|f_1(\tau, \omega)| \leqslant k e^{\varepsilon\tau} \leqslant \frac{k}{\Phi_\varepsilon^-} x_\varepsilon(\tau, \omega)$$

everywhere on $(0, \infty) \times \Omega$ (compare Theorem 21). Hence (compare (81))

$$|\hat{A}^n f(\tau, \omega)| \leqslant M^{n-4}(\varepsilon) \frac{k}{\Phi_\varepsilon^-} x_\varepsilon(\tau, \omega),$$

which means that

$$e^{-\varepsilon\tau} |\hat{A}^n f(\tau, \omega)| \leqslant k \frac{\Phi_\varepsilon^+}{\Phi_\varepsilon^-} M^{n-4}(\varepsilon), \quad n \geqslant 4, \quad (\tau, \omega) \in (0, \infty) \times \Omega.$$

Since $|\epsilon| < \lambda_0$, it follows that $0 < M(\epsilon) < M(\lambda_0) = 1$. Statement [i] is now self-evident.

The sum of the Neumann series

$$s(\tau, \omega) = \sum_{n=0}^{\infty} \hat{A}^n f(\tau, \omega)$$

can now be expressed as follows:

$$s(\tau, \omega) = x(\tau, \omega) + u(\tau, \omega),$$

$$x(\tau, \omega) = \sum_{n=0}^{3} \hat{A}^n f(\tau, \omega), \quad u(\tau, \omega) = \sum_{n=0}^{\infty} \hat{A}^n f_1(\tau, \omega), \quad (\tau, \omega) \in (0, \infty) \times \Omega.$$

Clearly, $x(\cdot) \in \mathcal{B}_\epsilon$ (compare Theorem 20). According to Theorem 17 [i], the functions $\hat{A}^n f(\cdot) = \hat{A}(\hat{A}^{n-1} f(\cdot))$ for $n \geq 1$ are strongly continuous with respect to τ on $(0, \infty)$. This in turn implies the continuity of the numerical functions $\hat{A}^n f(\tau, \omega)$, $n \geq 2$, with respect to (τ, ω) on $(0, \infty) \times \Omega$ (cf. the corollary of Theorem 17). From the uniform convergence of the Neumann series on sets of the form $\mathcal{F} \times \Omega$ which we established in [i], we conclude that $u(\tau, \omega)$ is continuous with respect to (τ, ω) on $(0, \infty) \times \Omega$. Here

$$e^{-\varepsilon\tau} |u(\tau, \omega)| \leqslant \sum_{n=0}^{\infty} e^{-\varepsilon\tau} |\hat{A}^n f_1(\tau, \omega)| \leqslant k \frac{\Phi_\varepsilon^+}{\Phi_\varepsilon^-} \sum_{n=4}^{\infty} M^{n-4}(\varepsilon)$$

$$= k \frac{\Phi_\varepsilon^+}{\Phi_\varepsilon^-} \cdot \frac{1}{1 - M(\varepsilon)}, \quad (\tau, \omega) \in (0, \infty) \times \Omega,$$

and statement [iii] is proved.

Since $u(\tau, \omega)$ is continuous with respect to ω on the sphere Ω, and since $x(\tau) \in L_2(\Omega)$ for $\tau > 0$, it follows that $s(\tau, \omega) = u(\tau, \omega) + x(\tau, \omega)$, being a function of $\omega \in \Omega$, is an element of $L_2(\Omega)$, $\tau > 0$. This implies that $s(\cdot) \in \mathcal{K}$. Clearly, the abstract function $u(\tau)$ with values in $L_2(\Omega)$ is strongly continuous, and therefore strongly measurable on $(0, \infty)$. Moreover, for all $\tau > 0$ we have

$$\| u(\tau) \| \leqslant 2\sqrt{\pi}\, k \frac{\Phi_\varepsilon^+}{\Phi_\varepsilon^-} \cdot \frac{1}{1 - M(\varepsilon)} e^{\varepsilon\tau}.$$

This means that $u(\cdot) \in \mathcal{B}_{\epsilon'}$ for all $\epsilon' \in (\epsilon, 1)$. At the same time $x(\cdot) \in \mathcal{B}_\epsilon \subset \mathcal{B}_{\epsilon'}$, $\epsilon' > \epsilon$. Thus $s(\cdot) \in \mathcal{B}_{\epsilon'}$ for every $\epsilon' \in (\epsilon, 1)$. Specifically, if $\epsilon_1 \in (\epsilon, \lambda_0)$, then $s(\cdot) \in \mathcal{B}_{\epsilon_1} \subset \mathcal{T}_{\lambda_0}$ (see the Definition in §9).

It remains for us to show that $s(\cdot)$ is a solution of the inhomogeneous Milne problem with the absolute term $f(\cdot)$: $s(\cdot) = \hat{A}s(\cdot) + f(\cdot)$. To accomplish this we need merely establish that

$$s(\tau, \omega) = \hat{A}s(\tau, \omega) + f(\tau, \omega)$$

at every point $(\tau, \omega) \in (0, \infty) \times \Omega$.

Let $N > 4$, $s(\cdot) = \sum_{n=0}^{N} \hat{A}^n f(\cdot) + r_N(\cdot)$. Here $r_N(\cdot) \in \overline{\mathcal{B}}_\epsilon r_N(\tau, \omega) = \sum_{n=N+1}^{\infty} \hat{A}^n f(\tau, \omega)$, so that

$$|r_N(\tau, \omega)| \leqslant \sum_{n=N+1}^{\infty} M^{n-4}(\varepsilon) \frac{k}{\Phi_\varepsilon^-} x_\varepsilon(\tau, \omega) = \frac{k}{\Phi_\varepsilon^-} \cdot \frac{M^{N-3}(\varepsilon)}{1 - M(\varepsilon)} x_\varepsilon(\tau, \omega)$$

everywhere on $(0, \infty) \times \Omega$. This means (compare (81)) that

$$|\hat{A}r_N(\tau, \omega)| \leqslant \hat{A}|r_N(\cdot)|(\tau, \omega) \leqslant \frac{k}{\Phi_\varepsilon^-}\frac{M^{N-3}(\varepsilon)}{1-M(\varepsilon)}\hat{A}x_\varepsilon(\tau, \omega)$$

$$\leqslant \frac{k}{\Phi_\varepsilon^-}\frac{M^{N-2}(\varepsilon)}{1-M(\varepsilon)}x_\varepsilon(\tau, \omega)$$

and

$$\lim_{N=\infty} \hat{A}r_N(\tau, \omega) = 0, \quad (\tau, \omega) \in (0, \infty) \times \Omega.$$

Now, taking the limit as $N \to \infty$ in the expression

$$\hat{A}s(\tau, \omega) + f(\tau, \omega) = \sum_{n=0}^{N+1} \hat{A}^n f(\tau, \omega) + \hat{A}r_N(\tau, \omega),$$

and making use of [i], we obtain

$$\hat{A}s(\tau, \omega) + f(\tau, \omega) = \sum_{n=0}^{\infty} \hat{A}^n f(\tau, \omega) = s(\tau, \omega), \ (\tau, \omega) \in (0, \infty) \times \Omega.$$

The lemma is proved.

In proving Lemma 26 we made essential use of the nonconservative condition $g_0 \neq 1$. We can rid ourselves of this assumption by appropriately limiting the rate of growth of the absolute term $f(\cdot)$ of equation (52) as $\tau \to \infty$. The result of this operation is contained in Lemma 27, which we preface with the following ancillary constructions.

Let $g_0 \leq 1$ and $\lambda \in (-1, 0]$. We set $\psi = \Phi_\lambda$, $s = M(\lambda)$. Then λ, s, and ψ satisfy the conditions of Lemma 23. Combining statements [iii] and [vi] of this lemma, we obtain

$$x_\lambda(\tau, \omega) = \sum_{\nu=0}^{\infty} M^{-\nu}(\lambda)\hat{A}^\nu p_\lambda(\tau, \omega),$$

so that (compare (81))

$$M^{-n}(\lambda)\,\hat{A}^n x_\lambda(\tau, \omega) = \sum_{\nu=n}^{\infty} M^{-\nu}(\lambda)\hat{A}^\nu p_\lambda(\tau, \omega),$$

$$(\tau, \omega) \in (0, \infty) \times \Omega, \ \lambda \in (-1, 0], \ n = 0, 1, 2, \ldots \tag{88}$$

Now let N be a nonnegative integer and let $\lambda \in (-1, 0)$. We choose an arbitrary $\lambda' \in (\lambda, 0] \cap [-\lambda_0, 0]$. Then $M(\lambda') < M(\lambda)$, $M(\lambda') \leq M(-\lambda_0) = 1$. We set $\theta = M(\lambda')/M(\lambda)$, so that $0 < \theta < 1$, and construct the expression

$$\beta_N^\lambda(\tau, \omega) = \sum_{n=N}^{\infty}\sum_{\nu=n}^{\infty} M^{n-\nu}(\lambda)\,\hat{A}^\nu p_\lambda(\tau, \omega).$$

Because the terms of this series are nonnegative we have

$$\beta_N^\lambda(\tau, \omega) = \sum_{\nu=N}^{\infty}\left[\sum_{n=N}^{\nu} M^{n-\nu}(\lambda)\right]\hat{A}^\nu p_\lambda(\tau, \omega).$$

But for all $N \geq 0$ and $\nu \geq N$ we immediately obtain

$$M^\nu(\lambda') \sum_{n=N}^{\nu} M^{n-\nu}(\lambda) = \sum_{n=N}^{\nu} \theta^{\nu-n} M^n(\lambda') \leqslant \sum_{n=N}^{\nu} \theta^{\nu-n} < \frac{1}{1-\theta}$$

and

$$\beta_N^\lambda(\tau, \omega) \leqslant \frac{1}{1-\theta} \sum_{\nu=N}^{\infty} M^{-\nu}(\lambda') \hat{A}^\nu p_\lambda(\tau, \omega). \tag{89}$$

Hence

$$\beta_N^\lambda(\tau, \omega) \leqslant \frac{\Phi_\lambda^+}{1-\theta} \sum_{\nu=N}^{\infty} g_0^{-\nu} \hat{A}^\nu p_0(\tau, \omega) \leqslant \frac{\Phi_\lambda^+}{1-\theta} < \infty, \quad (\tau, \omega) \in (0, \infty) \times \Omega.$$

As usual, let $\mathfrak{F}$ be a bounded closed portion of the halfline $(0, \infty)$. Making use of the latter estimate and Lemma 22, we find that

$$\lim_{N=\infty} \beta_N^\lambda(\tau, \omega) = 0$$

uniformly with respect to $(\tau, \omega) \in \mathfrak{F} \times \Omega$.

But $\beta_N^\lambda(\tau, \omega) = \Sigma_{n=N}^\infty \hat{A}^n x_\lambda(\tau, \omega)$ (cf. (88)), and we conclude:

If $\lambda \in (-1, 0)$, then the series

$$\sum_{n=0}^{\infty} \hat{A}^n x_\lambda(\tau, \omega) = z_\lambda(\tau, \omega). \tag{90}$$

converges uniformly with respect to all $(\tau, \omega) \in \mathfrak{F} \times \Omega$. Here (compare (89) and (88))

$$z_\lambda(\tau, \omega) = \beta_0^\lambda(\tau, \omega) \leqslant \frac{1}{1-\theta} \frac{\Phi_\lambda^+}{\Phi_{\lambda'}^-} \sum_{\nu=0}^{\infty} M^{-\nu}(\lambda') \hat{A}^\nu p_{\lambda'}(\tau, \omega) = \frac{1}{1-\theta} \frac{\Phi_\lambda^+}{\Phi_{\lambda'}^-} x_{\lambda'}(\tau, \omega)$$

and

$$0 \leqslant z_\lambda(\tau, \omega) \leqslant \frac{M(\lambda)}{M(\lambda) - M(\lambda')} \frac{\Phi_\lambda^+ \Phi_{\lambda'}^+}{\Phi_{\lambda'}^-} e^{\lambda'\tau}, \quad (\tau, \omega) \in (0, \infty) \times \Omega,$$

$$\lambda \in (-1, 0), \quad \lambda' \in (\lambda, 0] \cap [-\lambda_0, 0]. \tag{91}$$

All of the functions $\hat{A}^n x_\lambda(\tau, \omega)$, $n \geq 0$, are continuous with respect to (τ, ω) (Theorem 17). Because the series (90) converges uniformly, this implies that the $\beta_N^\lambda(\tau, \omega)$ are also continuous on $(0, \infty) \times \Omega$, $N \geq 0$. This fact together with the inequality $\beta_N^\lambda(\tau, \omega) \leq z_\lambda(\tau, \omega)$ and estimate (91) ensures the inclusion $\beta_N^\lambda(\cdot) \in \overline{\mathfrak{B}}$, $N = 0, 1, 2, \cdots$. This means that the expressions $\hat{A}\beta_N^\lambda(\cdot)$ have meaning. It follows from (89) that

$$\hat{A}\beta_N^\lambda(\tau, \omega) \leqslant \frac{\Phi_\lambda^+}{1-\theta} \hat{A}\left[\sum_{\nu=N}^{\infty} g_0^{-\nu} \hat{A}^\nu p_0(\cdot)\right](\tau, \omega).$$

This and Lemma 22 [iii] imply that

$$\lim_{N=\infty} \hat{A}\beta_N^\lambda(\tau, \omega) = 0 \tag{92}$$

uniformly with respect to $(\tau, \omega) \in \mathfrak{F} \times \Omega$.

Lemma 27. *Let $\epsilon \in (-1, 0)$, and let $\mathfrak{F}$ be a bounded closed portion of the halfline $(0, \infty)$; let $f(\cdot) \in \mathfrak{B}_\epsilon$. Then*

[i] *The series*

$$\sum_{n=0}^{\infty} \hat{A}^n f(\tau, \omega) = s(\tau, \omega)$$

converges uniformly with respect to all $(\tau, \omega) \in \mathfrak{F} \times \Omega$ at every point $(\tau, \omega) \in (0, \infty) \times \Omega$. Being an abstract function of the variable $\tau \in (0, \infty)$, the sum $s(\cdot)$ of this series has the following properties:

[ii] *$s(\cdot) \in \mathfrak{T}_{\lambda_0}$, and $s(\cdot) \in \mathfrak{B}_{\epsilon'}$ for every $\epsilon' \in (\max\{-\lambda_0, \epsilon\}, 1)$.*

[iii] *$s(\cdot) = x(\cdot) + u(\cdot)$, where $x(\cdot) = \Sigma_{n=0}^3 \hat{A}^n f(\cdot) \in \mathfrak{B}_\epsilon$, and the numerical function $u(\tau, \omega)$ is continuous with respect to the pair of variables (τ, ω) on $(0, \infty) \times \Omega$, where*

$$\sup_{(\tau, \omega) \in (0, \infty) \times \Omega} e^{-\epsilon'\tau} |u(\tau, \omega)| < \infty$$

for every $\epsilon' \in (\epsilon, 0] \cap [-\lambda_0, 0]$.

[iv] *$s(\cdot) = \hat{A}s(\cdot) + f(\cdot)$.*

Proof. As in our proof of Lemma 26, we set $x(\cdot) = \Sigma_{\nu=0}^3 \hat{A}^\nu f(\cdot)$, and $f_1(\cdot) = \hat{A}^4 f(\cdot)$. Then $x(\cdot) \in \mathfrak{B}_\epsilon$ and everywhere on $(0, \infty) \times \Omega$ we have

$$|f_1(\tau, \omega)| \leqslant kx_\varepsilon(\tau, \omega), \quad 0 < k = \text{const} < \infty;$$
$$|\hat{A}^n f_1(\tau, \omega)| \leqslant k\hat{A}^n x_\varepsilon(\tau, \omega), \quad n = 0, 1, 2, \ldots.$$

This and (90) (with $\lambda = \epsilon \in (-1, 0)$) imply statement [i] of the lemma. In particular, the series defining the function $u(\tau, \omega)$, i.e.

$$u(\tau, \omega) \equiv s(\tau, \omega) - x(\tau, \omega) = \sum_{n=4}^{\infty} \hat{A}^n f(\tau, \omega) = \sum_{n=0}^{\infty} \hat{A}^n f_1(\tau, \omega)$$

converges uniformly with respect to $(\tau, \omega) \in \mathfrak{F} \times \Omega$. Since $\mathfrak{F}$ is continuous, and since $\hat{A}^n f(\tau, \omega)$, $n \geq 2$, is continuous with respect to (τ, ω) on $(0, \infty) \times \Omega$ (Theorem 17), the function $u(\tau, \omega)$ is also continuous on $(0, \infty) \times \Omega$. Comparing the inequality

$$|u(\tau, \omega)| \leqslant \sum_{n=0}^{\infty} |\hat{A}^n f_1(\tau, \omega)| \leqslant k \sum_{n=0}^{\infty} \hat{A}^n x_\varepsilon(\tau, \omega) = kz_\varepsilon(\tau, \omega)$$

with estimate (91), we arrive at [iii].

Now let $\max\{-\lambda_0, \epsilon\} < \epsilon' < 1$. Clearly, $x(\cdot) \in \mathcal{B}_\epsilon \subset \mathcal{B}_{\epsilon'}$. Let us consider $u(\tau, \omega)$ as an abstract function $u(\cdot)$ of the variable $\tau \in (0, \infty)$ with values in $L_2(\Omega)$. It is clear that $u(\tau)$ is strongly continuous, and therefore strongly measurable, on $(0, \infty)$. Since $0 \in (-\epsilon, 0] \cap [-\lambda_0, 0]$, it follows by [iii] that $\sup_{\tau>0} \|u(\tau)\| < \infty$. This implies immediately that $u(\cdot) \in \mathcal{B}_{\epsilon'}$ for $\epsilon' > 0$. Let us assume that $\epsilon' \leq 0$. The inequalities $\epsilon' > -\lambda_0$ and $\epsilon' > \epsilon$ enable us to choose a number ϵ'' such that $\epsilon'' > -\lambda_0$, $0 \geq \epsilon' > \epsilon'' > \epsilon$. Thus $\epsilon'' \in (\epsilon, 0] \cap [-\lambda_0, 0]$ and $\|u(\tau)\| \leq ae^{\epsilon''\tau}$ for all $\tau > 0$ (compare [iii], $0 < a = \text{const} < \infty$). Recalling that $\epsilon' > \epsilon''$, we obtain the inclusion $u(\cdot) \in \mathcal{B}_{\epsilon'}$ in the case $\epsilon' \leq 0$ as well.

Thus for every $\epsilon' \in (\max\{-\lambda_0, \epsilon\}, 1)$ we have

$$s(\cdot) = x(\cdot) + u(\cdot) \in \mathcal{B}_{\varepsilon'} + \mathcal{B}_{\varepsilon'} \subset \mathcal{B}_{\varepsilon'}.$$

We note now that $x(\cdot) \in \mathcal{B}_\epsilon \subset \mathcal{B}_0$, and, as shown above, that $u(\cdot) \in \mathcal{B}_1$, $\sup_{\tau>0} \|u(\tau)\| < \infty$. By the definition of the class $\mathcal{T}_{\lambda_0}$ (§9), this implies that $s(\cdot) = x(\cdot) + u(\cdot) \in \mathcal{T}_{\lambda_0}$. Statement [ii] is proved.

For every natural number $N \geq 3$ we set

$$r_N(\tau, \omega) = s(\tau, \omega) - \sum_{n=0}^{N} \hat{A}^n f(\tau, \omega) = \sum_{n=N+1}^{\infty} \hat{A}^n f(\tau, \omega)$$

$$= \sum_{n=N-3}^{\infty} \hat{A}^n f_1(\tau, \omega), \quad (\tau, \omega) \in (0, \infty) \times \Omega.$$

It is clear that $r_N(\cdot) \in \overline{\mathcal{B}}$ and

$$|r_N(\tau, \omega)| \leqslant k \sum_{n=N-3}^{\infty} \hat{A}^n x_\varepsilon(\tau, \omega) = k\beta^\varepsilon_{N-3}(\tau, \omega),$$

$$|\hat{A} r_N(\tau, \omega)| \leqslant k\hat{A}\beta^\varepsilon_{N-3}(\tau, \omega), \quad (\tau, \omega) \in (0, \infty) \times \Omega.$$

From this and (92) we infer that

$$\lim_{N=\infty} \hat{A} r_N(\tau, \omega) = 0$$

everywhere on $(0, \infty) \times \Omega$.

Now let us take the limit as $N \to \infty$ in the expression

$$\hat{A}s(\tau, \omega) + f(\tau, \omega) = \sum_{n=0}^{N+1} \hat{A}^n f(\tau, \omega) + \hat{A} r_N(\tau, \omega).$$

This yields

$$\hat{A}s(\tau, \omega) + f(\tau, \omega) = \sum_{n=0}^{\infty} \hat{A}^n f(\tau, \omega) = s(\tau, \omega),$$

which implies [iv]. The lemma is proved.

Omitting the details, we can combine Lemmas 25–27 into the following theorem.

Theorem 26. *Let* $-1 < \epsilon < \lambda_0$ *and* $f(\cdot) \in \mathfrak{B}_\epsilon$. *Then the class* $\mathfrak{T}_{\lambda_0}$ *contains exactly one solution* $x(\cdot) = s(\cdot)$ *of the equation*

$$x(\cdot) = \hat{A}\, x(\cdot) + f(\cdot).$$

This solution $s(\cdot) \in \mathfrak{T}_{\lambda_0}$ *is the sum of the corresponding Neumann series which converges everywhere on* $(0, \infty) \times \Omega$:

$$s(\tau, \omega) = \sum_{n=0}^{\infty} \hat{A}^n f(\tau, \omega).$$

If $\mathfrak{F}$ *is a bounded closed portion of the halfline* $(0, \infty)$, *then the Neumann series* $\sum_{n=0}^{\infty} \hat{A}^n f(\tau, \omega)$ *converges uniformly with respect to all* $(\tau, \omega) \in \mathfrak{F} \times \Omega$.

Proof. The uniqueness of the solution of the equation $x(\cdot) = \hat{A}x(\cdot) + f(\cdot)$ in the class $\mathfrak{T}_{\lambda_0}$ is ensured by Lemma 25. If $-1 < \epsilon < 0$, then the remaining statements of the theorem are contained in Lemma 27. On the other hand, if $\epsilon > 0$, then $\lambda_0 > 1$, which means that $g_0 < 1$. This implies fulfillment of all the conditions of Lemma 26, from which the theorem follows.

Like Lemma 25, this theorem can be strengthened by means of the results of §11.

We shall now consider the special case of equation (52) mentioned at the beginning of this section. We shall assume that the material filling the halfspace $\tau > 0$ contains no radiation sources, and that the radiation incident on the boundary $\tau = 0$ is distributed according to the law $B(\omega)$.

Definition. Problem I consists in solving the equation $x(\cdot) = \hat{A}x(\cdot) + f(\cdot)$, where

$$f(\tau, \omega) = B(\omega)\, e^{-\tau/\omega n}, \quad (\tau, \omega) \in (0, \infty) \times \Omega,$$

in which the function $B(\omega)$ satisfies the restrictions

$$B \in L_2(\Omega); \quad B(\omega) = 0 \quad \text{for } \omega n \leqslant 0.$$

Problem I is posed once the function $B(\omega)$ satisfying the conditions of this definition has been prescribed. The homogeneous Milne problem is a special case of Problem I corresponding to the assumption that $B(\omega) \equiv 0$ on Ω.

The absolute term $f(\cdot)$ in Problem I has the following important properties:

$$f(\cdot) \in \mathcal{B}_\varepsilon \quad \text{for all} \quad \varepsilon \in (-1, 1), \tag{93}$$

$$\lim_{\tau=\infty} f(\tau, \omega) = B(\omega) \tag{94}$$

for every $\omega \in \Omega$.

$$\lim_{\tau=0+0} \|f(\tau) - B\| = 0. \tag{95}$$

$$\sup_{\tau>0} e^{-\varepsilon\tau} \| \hat{A}^n f(\tau) \| < \infty, \quad n = 0, 1, 2, \ldots, \varepsilon \in (-1, 1). \tag{96}$$

In fact, $f(\tau) \in L_2(\Omega)$, $\|f(\tau)\| \leq \|B\| e^{-\tau}$ for every $\tau > 0$. On the other hand, if $\tau_{1,2} > 0$, then

$$\|f(\tau_1) - f(\tau_2)\|^2 = \int_{\omega n > 0} |B(\omega)|^2 |e^{-\tau_1/\omega n} - e^{-\tau_2/\omega n}|^2 d\omega.$$

The integrand is bounded by the summable function $|B(\omega)|^2$. Hence, by the Lebesgue theorem on dominated convergence, $\lim_{\tau_2=\tau_1} \|f(\tau_1) - f(\tau_2)\| = 0$. This means that $f(\tau)$ in Problem I is always strongly convergent with respect to τ on $(0, \infty)$. Inclusion (93) is now self-evident.

Limit relation (94) follows directly from the definition of $f(\cdot)$. (95) follows from (94) and the Lebesgue theorem on dominated convergence.

We choose an $\epsilon \in (-1, 1)$ and note that

$$| \hat{A}^0 f(\tau, \omega) | = |f(\tau, \omega)| \leqslant |B(\omega)| e^{-\tau} \leqslant |B(\omega)| e^{\varepsilon\tau}$$

everywhere on $(0, \infty) \times \Omega$. Hence (compare (76) and (81))

$$| \hat{A} f(\tau, \omega) | \leqslant \frac{e^{\varepsilon\tau}}{1 + \varepsilon\omega n} \hat{g} |B| (\omega) \leqslant \frac{\sqrt{2\pi}\,[g]\,\|B\|}{(1 - |\varepsilon|)\,\Phi_\varepsilon^-} x_\varepsilon(\tau, \omega),$$

$$| \hat{A}^n f(\tau, \omega) | \leqslant \frac{\sqrt{2\pi}\,[g]\,\|B\|}{(1 - |\varepsilon|)\,\Phi_\varepsilon^-} \hat{A}^{n-1} x_\varepsilon(\tau, \omega) \leqslant M^{n-1}(\varepsilon) \frac{\sqrt{2\pi}\,[g]\,\|B\|}{(1 - |\varepsilon|)\,\Phi_\varepsilon^-} \Phi_\varepsilon(\omega) e^{\varepsilon\tau},$$

$$(\tau, \omega) \in (0, \infty) \times \Omega, \quad n = 1, 2, \ldots.$$

Thus for all $\tau > 0$ we have

$$e^{-\varepsilon\tau} \| \hat{A}^0 f(\tau) \| \leqslant \|B\| = \text{const} < \infty,$$

$$e^{-\varepsilon\tau} \| \hat{A}^n f(\tau) \| \leqslant M^{n-1}(\varepsilon) \frac{\sqrt{2\pi}\,[g]\,\|B\|}{(1 - |\varepsilon|)\,\Phi_\varepsilon^-} \|\Phi_\varepsilon\| = \text{const} < \infty,$$

$$n = 1, 2, \ldots,$$

and (96) is proved.

Inclusion (93) and Theorem 26 guarantee the existence of solutions of Problem I, of which exactly one belongs to the class $\mathcal{T}_{\lambda_0}$. From now on we shall always denote this solution by the symbol $s_I(\cdot)$:

$$s_I(\cdot) = \hat{A} s_I(\cdot) + f(\cdot), \quad s_I(\cdot) \in \mathcal{T}_{\lambda_0}. \tag{97}$$

We choose an $\epsilon \in (-1, -\lambda_0)$. Then, in accordance with (93), $f(\cdot) \in \mathcal{B}_\epsilon$. Making use of Lemma 27 [ii],[iii], we find that

$$s_I(\cdot) \in \mathcal{B}_{\varepsilon'} \quad \text{for every} \quad \varepsilon' \in (-\lambda_0, 1), \tag{98}$$

and $s_I(\cdot) = \Sigma_{n=0}^{3} \hat{A}^n f(\cdot) + u(\cdot)$, where $\sup_{\tau>0} \|u(\tau)\| < \infty$.

Together with (96), this result implies that

$$\sup_{\tau>0} \|s_I(\tau)\| < \infty. \tag{99}$$

Formula (88) yields an interesting example of a solution of Problem I. Let us set $B(\omega) = \eta(\omega\mathbf{n})\Phi_{-\lambda_0}(\omega)$, $\omega \in \Omega$. Clearly, $B \in L_2(\Omega)$, $B(\omega) = 0$ for $\omega\mathbf{n} \le 0$. The absolute term $f(\cdot)$ in Problem I becomes $f(\cdot) = p_{-\lambda_0}(\cdot)$.

Recalling that $-\lambda_0 \in (-1, 0]$, we find from (88) that

$$s_I(\tau, \omega) = \sum_{\nu=0}^{\infty} \hat{A}^\nu p_{-\lambda_0}(\tau, \omega) = x_{-\lambda_0}(\tau, \omega) = e^{-\lambda_0\tau}\Phi_{-\lambda_0}(\omega)$$

and that

$$s_I(\cdot) = x_{-\lambda_0}(\cdot).$$

This example shows that the inclusion $s_I(\cdot) \in \mathcal{B}_{\epsilon'}$ generally does not hold for $\epsilon' \le -\lambda_0$.

Now let $x(\cdot) \in \overline{\mathcal{B}}$ be some solution of Problem I. We define the function $z^0(\omega)$, $\omega \in \Omega$, in accordance with formula (63). By Theorem 18 we know that $z^0 \in L_2(\Omega)$, and

$$\lim_{\tau=0+0} \|\hat{A}x(\tau) - z^0\| = 0.$$

Thus the right side of the expression

$$x(\tau) = \hat{A}x(\tau) + f(\tau), \quad \tau > 0,$$

has a limit (compare (95)) in the metric of $L_2(\Omega)$ as $\tau \to 0 + 0$. This means that $x(\tau)$ also has a limit as $\tau \to 0 + 0$. We denote this limit by $x(0)$,

$$\lim_{\tau=0+0} x(\tau) = x(0) \in L_2(\Omega).$$

Let us set

$$z^{00}(\omega) = \begin{cases} z^0(\omega) & \text{for } \omega\mathbf{n} \ne 0, \\ x^*_\omega(\hat{g}x(0)) & \text{for } \omega\mathbf{n} = 0. \end{cases}$$

$z^{00}(\omega)$ differs from $z^0(\omega)$ only on a set of zero measure. Hence, being an element of the space $L_2(\Omega)$, $z^{00} = z^0$. The properties of $z^{00}(\omega)$ are described in Theorem 19. It is clear that $x(0) = z^{00} + B$, which means that

$$x(0,\omega)=\begin{cases}B(\omega) & \text{for } \omega\mathbf{n}>0,\\ z^{00}(\omega) & \text{for } \omega\mathbf{n}\leqslant 0.\end{cases} \tag{100}$$

Let $\tau > 0$. For all $\omega\mathbf{n} \le 0$ we have

$$|x(\tau,\omega)-x(0,\omega)|=|\hat{A}x(\tau,\omega)-z^{00}(\omega)|.$$

Theorem 19 now implies that

$$\lim_{\tau=0+0} x(\tau,\omega)=x(0,\omega),$$

uniformly with respect to all $\omega \in \Omega^- \equiv \{\omega' | \omega' \in \Omega,\ \omega'\mathbf{n} \le 0\}$, and that the limit function $x(0,\omega)$ is continuous on Ω^-.

Theorem 27. *Let $\epsilon \in (-1, 1)$, let $x(\cdot)$ be a solution of Problem* I, *and let $x(\cdot) \in \mathfrak{B}_\epsilon$. Then*

[i] *There exists a function $x(0) \in L_2(\Omega)$ such that*

a) $x(0,\omega)$ *is continuous on the closed hemisphere Ω^-; $x(0,\omega) = B(\omega)$ for $\omega\mathbf{n} > 0$.*

b) $\lim_{\tau=0+0} x(\tau,\omega) = x(0,\omega)$ *uniformly with respect to all $\omega \in \Omega^-$.*

c) $\lim_{\tau=0+0} \|x(\tau) - x(0)\| = 0$.

[ii] $\sup_{\tau>0} e^{-\epsilon\tau}\|x(\tau)\| < \infty$.

Proof. Statement [i] is proved above. It is clear that for $\tau > 0$ we have

$$x(\tau)=\hat{A}^3x(\tau)+f'(\tau),\ f'(\tau)=\sum_{n=0}^{2}\hat{A}^nf(\tau).$$

According to Theorem 21, there exists a number k_0, $0 < k_0 < \infty$, such that $\|\hat{A}^3x(\tau)\| \le k_0 e^{\epsilon\tau}$ for all $\tau > 0$. On the other hand, according to (26), $\sup_{\tau>0} e^{-\epsilon\tau}\|f'(\tau)\| < \infty$, which implies the theorem.

§11. Problem I. The Fourier method

We already know that the homogeneous Milne problem admits of nontrivial solutions. In §9 we constructed a linearly independent system $\mathfrak{X}$ of such solutions and investigated some of its properties. Here we shall show that every solution of the homogeneous Milne problem which belongs to the class $\overline{\mathfrak{B}}$ is a linear combination of a finite number of elements of the system $\mathfrak{X}$. In order to arrive at this result, we shall make use of Laplace transformation of abstract functions from the classes $\mathfrak{B}_\epsilon$. It will be more convenient for us to consider not the homogeneous Milne problem as such, but rather the more general Problem I posed in the preceding section. It turns out that the solutions of Problem I admit of simple asymptotic expansions for $\tau \to \infty$ (Theorem 28). The chief result of the present section, Theorem 30, is obtained by comparing the asymptotic form of the solution of the Milne problem with functions from the system $\mathfrak{X}$.

In speaking of Problem I we shall always assume that we are dealing with one of its specific variants determined by prescribing the function $B \in L_2(\Omega)$ (see the Definition in §10). Much of the notation introduced in the earlier sections (e.g. the sectioned plane Z_0, the set $\mathfrak{N}$, et al.) will be used without further explanation.

Let ϵ be a real number, and let $x(\cdot) \in \mathfrak{B}_\epsilon$ and $\operatorname{Re} k > \epsilon$. The abstract function $e^{-k\tau}x(\tau)$ is strongly measurable on $(0, \infty)$, and $\|e^{-k\tau}x(\tau)\| \le e^{-\epsilon\tau}\|x(\tau)\|$. This means that the Laplace-Bochner integral

$$\tilde{\mathfrak{X}}(k) = \int_0^\infty e^{-k\tau}x(\tau)\,d\tau \in L_2(\Omega), \quad x(\cdot) \in \mathfrak{B}_\varepsilon, \quad \operatorname{Re} k > \varepsilon, \tag{101}$$

is defined. Let $u \in L_2(\Omega)$. By formula (53),

$$(\tilde{\mathfrak{X}}(k), u) = \int_0^\infty e^{-k\tau}(x(\tau), u)\,d\tau. \tag{101'}$$

Making use of the summability on $(0, \infty)$ of the numerical function $e^{-\epsilon\tau}\|x(\tau)\|$, we can easily establish in the ordinary way that the function $(\tilde{\mathfrak{X}}(k), u)$ is holomorphic with respect to the variable k in the halfplane $\{k \mid \operatorname{Re} k > \epsilon\}$. Since u is an arbitrary element of $L_2(\Omega)$, this immediately implies [35] that $\tilde{\mathfrak{X}}(k)$ is an abstract holomorphic function of k in the same halfplane.

All we shall need is the following lemma on the inversion of Laplace transform (101).

Lemma 28. *Let ϵ be a real number, and let $x(\cdot) \in \mathfrak{B}_\epsilon$ and $\sup_{\tau>0} e^{-\epsilon\tau}\|x(\tau)\| < \infty$. If $u \in L_2(\Omega)$, if $\gamma > \epsilon$, and if $\{a_n \mid n = 1, 2, \dots\}$ is an arbitrary sequence of positive numbers, $\lim_{n=\infty} a_n = \infty$, then*

$$e^{-\gamma\tau}(x(\tau), u) = \underset{n=\infty}{\mathrm{l.\,i.\,m.}}\ \frac{1}{2\pi i}\int_{\gamma - ia_n}^{\gamma + ia_n} e^{(k-\gamma)\tau}(\tilde{\mathfrak{X}}(k), u)\,dk.$$

(By the symbol l.i.m. *in the right side we mean the limit of the sequence of functions of the variable $\tau \in (0, \infty)$ in the metric of $L_2(0, \infty)$.)*

To prove this we set

$$q(\tau) = \begin{cases} (x(\tau), u)\,e^{-\gamma\tau} & \text{for } \tau > 0, \\ 0 & \text{for } \tau \leqslant 0. \end{cases}$$

Then $|q(\tau)| \le b\|u\|e^{-e(\gamma-\epsilon)\tau}$, $\tau > 0$, $0 < b = \text{const} < \infty$, which means that $|q(\tau)|$ and $|q(\tau)|^2$ are summable on $(-\infty, \infty)$. We denote by $\tilde{q}(s)$ the Fourier transform of the function $q(\tau)$,

$$\tilde{q}(s) = \frac{1}{\sqrt{2\pi}}\int_{-\infty}^{\infty} e^{i\tau s}q(\tau)\,d\tau = \frac{1}{\sqrt{2\pi}}\int_0^\infty e^{-(\gamma - is)\tau}(x(\tau), u)\,d\tau = \frac{1}{\sqrt{2\pi}}(\tilde{\mathfrak{X}}(\gamma - is), u).$$

According to the Plancherel theory of Fourier transformations in $L_2(-\infty, \infty)$ [39], we have

$$q(\tau) = \underset{n=\infty}{\mathrm{l.\,i.\,m.}} \frac{1}{\sqrt{2\pi}} \int_{-a_n}^{a_n} e^{-i\tau s} \tilde{q}(s)\, ds = \underset{n=\infty}{\mathrm{l.\,i.\,m.}} \frac{1}{2\pi i} \int_{\gamma - ia_n}^{\gamma + ia_n} e^{(k-\gamma)\tau} (\tilde{x}(k), u)\, dk,$$

Q.E.D.

Lemma 29. *Let $\epsilon \in (-1, 1)$, $x(\cdot) \in \mathcal{B}_\epsilon$ and $z(\cdot) = \hat{A}x(\cdot)$. For $k \in Z_0$ and $\mathrm{Re}\, k > \epsilon$, the Laplace images $\tilde{z}(k)$ and $\tilde{x}(k)$ of the functions $z(\cdot)$ and $x(\cdot)$ are related by the expression*

$$(1 + k\omega n)\, \tilde{z}\,(k) = \hat{g}\tilde{x}\,(k) + (\omega n)\, z^0,$$

where z^0 is the function from $L_2(\Omega)$ defined by formula (63).

We begin our proof of the lemma with the following remark. We choose an orthonormal basis σ of the space R_3 with the third unit vector n. Let $y(\cdot)$ be an abstract function of the variable $\tau \in (0, \infty)$ with values in $C(\Omega)$ which is strongly measurable on $(0, \infty)$. The numerical function $y(\tau, \psi_\sigma(\mu, \phi))$ of the three variables (τ, μ, ϕ) is then measurable on $(0, \infty) \times [-1, 1] \times [0, 2\pi]$.

In order to see this we assume first that

$$y(\tau) = \begin{cases} y_0 & \text{for } \tau \in \mathcal{E}, \\ \vartheta & \text{for } \tau \in (0, \infty) \setminus \mathcal{E}, \end{cases}$$

where $y_0 \in C(\Omega)$ and where $\mathcal{E}$ is a measurable portion of the halfline $(0, \infty)$.

If a is a real number, and if

$$X_a \equiv \{(\tau, \mu, \varphi) \mid (\tau, \mu, \varphi) \in (0, \infty) \times [-1, 1] \times [0, 2\pi],\ y(\tau, \psi_\sigma(\mu, \varphi)) > a\},$$

then for $a \geq 0$ we have

$$X_a = \mathcal{E} \times \{(\mu, \varphi) \mid (\mu, \varphi) \in \Delta,\ y_0(\psi_\sigma(\mu, \varphi)) > a\},$$

and, for $a < 0$,

$$X_a = [\mathcal{E} \times \{(\mu, \varphi) \mid (\mu, \varphi) \in \Delta,\ y_0(\psi_\sigma(\mu, \varphi)) > a\}] \cup [((0, \infty) \setminus \mathcal{E}) \times \Delta].$$

The function $y_0(\psi_\sigma(\mu, \phi))$ is continuous with respect to (μ, ϕ) on Δ. This implies the measurability of X_a, and therefore the measurability of the functions $y(\tau, \psi_\sigma(\mu, \phi))$.

If $y(\cdot)$ is a strongly measurable function which assumes a countable number of values, then the above result immediately implies the measurability of $y(\tau, \psi_\sigma(\mu, \phi))$.

Finally, let $y(\cdot)$ be an arbitrary strongly measurable function. There then exists [35] a sequence $\{y_n(\cdot)\}$ of strongly measurable countable-valued functions such that

$$\lim_{n=\infty} \| y_n(\tau) - y(\tau) \|_{C(\Omega)} = 0$$

almost everywhere on $(0, \infty)$. By the definition of the metric of $C(\Omega)$ this implies that

$$\lim_{n=\infty} y_n(\tau, \psi_\sigma(\mu, \varphi)) = y(\tau, \psi_\sigma(\mu, \varphi))$$

for almost all (τ, μ, ϕ). The fact that the $y_n(\cdot)$ are countable-valued implies that all of the $y_n(\tau, \psi_\sigma(\mu, \phi))$ are measurable with respect to (τ, μ, ϕ). The measurability of $y(\tau, \psi_\sigma(\mu, \phi))$ is now self-evident.

Let us return to the conditions of the lemma and set $y(\tau) = \hat{g}x(\tau)$, $\tau > 0$. By Lemma 19, $y(\cdot)$ is strongly measurable as an abstract function with values in $C(\Omega)$. This means that the numerical function $y(\tau, \psi_\sigma(\mu, \phi))$ is measurable with respect to (τ, μ, ϕ) on $(0, \infty) \times [-1, 1] \times [0, 2\pi]$.

We choose an arbitrary vector $u \in L_2(\Omega)$ and let $k \in Z_0$, $\operatorname{Re} k > \epsilon$. Because the integral $\int_0^\infty e^{-\epsilon\tau} \| z(\tau) \| d\tau$ is finite (compare Theorem 20), the Laplace image $\tilde{z}(k) = \int_0^\infty e^{-k\tau} z(\tau)\, d\tau$ is defined, and $(\tilde{z}(k), u) = \int_0^\infty e^{-k\tau}(z(\tau), u) d\tau$. But $z(\tau) = z_+(\tau) + z_-(\tau) + z_0(\tau)$ (compare (61), (57)–(59)). The finiteness of the integrals $\int_0^\infty e^{-\epsilon\tau} \| z_\pm(\tau) \| d\tau$ was established in our proof of Theorem 20. This implies the existence of the integrals $I_\pm = \int_0^\infty e^{-k\tau}(z_\pm(\tau), u) d\tau$ and the equation $(\tilde{z}(k), u) = I_+ + I_-$. We set

$$F_+(\tau, \rho, \mu, \varphi) = e^{-k\tau} \overline{u(\psi_\sigma(\mu, \varphi))} \cdot \frac{1}{\mu} e^{-\frac{\tau - \rho}{\mu}} y(\rho, \psi_\sigma(\mu, \varphi)),$$

$$\tau \in (0, \infty),\ \rho \in (0, \tau),\ \mu \in (0, 1],\ \varphi \in [0, 2\pi].$$

It is clear that F_+ is measurable over the collection of variables (τ, ρ, μ, ϕ) and (compare Lemma 1) that

$$I_+ = \int_0^\infty d\tau \int_0^1 \int_0^{2\pi} d\mu\, d\varphi \int_0^\tau F_+(\tau, \rho, \mu, \varphi)\, d\rho.$$

It is not difficult to verify the summability of F_+ in the indicated range of the independent variables. In fact, making use of the Fubini theorem for nonnegative measurable functions, we obtain

$$\int | F_+(\tau, \rho, \mu, \varphi) | d\tau\, d\rho\, d\mu\, d\varphi = \int_0^1 \int_0^{2\pi} d\mu\, d\varphi \int_0^\infty d\rho \int_\rho^\infty | F_+(\tau, \rho, \mu, \varphi) | d\tau$$

$$\leqslant \sqrt{2\pi}\, [g] \int_0^1 \int_0^{2\pi} \frac{1}{\mu} | u(\psi_\sigma(\mu, \varphi)) | d\mu\, d\varphi \int_0^\infty \| x(\rho) \| d\rho \int_\rho^\infty e^{-\epsilon\tau - \frac{\tau - \rho}{\mu}} d\tau =$$

$$= \sqrt{2\pi}\,[g]\int_0^1\int_0^{2\pi} |u(\psi_\sigma(\mu,\varphi))|\,d\mu\,d\varphi \int_0^\infty \frac{\|x(\rho)\|}{1+\varepsilon\mu} e^{-\varepsilon\rho}\,d\rho$$

$$\leqslant \sqrt{2\pi}\,[g]\,(1-|\varepsilon|)^{-1}(|u|,1)\int_0^\infty \|x(\rho)\|\,e^{-\varepsilon\rho}\,d\rho < \infty.$$

The summability of F_+ enables us to change the order of integration in I_+:

$$I_+ = \int_0^1\int_0^{2\pi} \frac{1}{\mu}\,\overline{u(\psi_\sigma(\mu,\varphi))}\,d\mu\,d\varphi \int_0^\infty d\rho \int_\rho^\infty e^{-k\tau-\frac{\tau-\rho}{\mu}}\, y(\rho,\psi_\sigma(\mu,\varphi))\,d\tau$$

$$= \int_0^1\int_0^{2\pi} \frac{1}{1+k\mu}\,\overline{u(\psi_\sigma(\mu,\varphi))}\,d\mu\,d\varphi \int_0^\infty y(\rho,\psi_\sigma(\mu,\varphi))\,e^{-k\rho}\,d\rho$$

$$= \int_0^\infty e^{-k\rho}\left[\int_0^1\int_0^{2\pi} \frac{y(\rho,\psi_\sigma(\mu,\varphi))}{1+k\mu}\,\overline{u(\psi_\sigma(\mu,\varphi))}\,d\mu\,d\varphi\right]d\rho$$

$$= \int_0^\infty e^{-k\rho}\left(\frac{y(\rho)}{1+k\omega n},\ \eta(\omega n)\,u\right)d\rho.$$

The change in the order of integration in I_- can be justified in a similar way:

$$I_- = \int_0^\infty e^{-k\tau}\,d\tau \int_{-1}^0\int_0^{2\pi} d\mu\,d\varphi\,\overline{u(\psi_\sigma(\mu,\varphi))}\,\frac{1}{|\mu|}\int_\tau^\infty e^{-\frac{\rho-\tau}{|\mu|}}\, y(\rho,\psi_\sigma(\mu,\varphi))\,d\rho$$

$$= \int_{-1}^0\int_0^{2\pi} \frac{1}{|\mu|}\,\overline{u(\psi_\sigma(\mu,\varphi))}\,d\mu\,d\varphi \int_0^\infty d\rho \int_0^\rho e^{-k\tau-\frac{\rho-\tau}{|\mu|}}\, y(\rho,\psi_\sigma(\mu,\varphi))\,d\tau$$

$$= \int_{-1}^0\int_0^{2\pi} (1-k|\mu|)^{-1}\,\overline{u(\psi_\sigma(\mu,\varphi))}\,d\mu\,d\varphi \int_0^\infty \left(e^{-k\rho}-e^{-\frac{\rho}{|\mu|}}\right) y(\rho,\psi_\sigma(\mu,\varphi))\,d\rho$$

$$= \int_0^\infty e^{-k\rho}\left(\frac{y(\rho)}{1+k\omega n},\ \eta(-\omega n)\,u\right)d\rho - \left(\frac{|\omega n|}{1+k\omega n}\,z^0,\,u\right).$$

Thus

$$(\tilde{z}(k),u) = I_+ + I_- = \int_0^\infty e^{-k\rho}\left(\frac{y(\rho)}{1+k\omega n},\ u\right)d\rho - \left(\frac{|\omega n|}{1+k\omega n}\,z^0,\ u\right).$$

We now note that if $e^{-k\rho}y(\rho)$ is considered as an abstract function with values in $L_2(\Omega)$, it turns out to be B-summable on $(0,\infty)$. In fact, $(y(\rho),v) = (x(\rho),\hat{g}v)$ is a measurable function of ρ for every $v\in L_2(\Omega)$ (since $x(\rho)$ is strongly measurable). This means that $y(\rho)$ and $e^{-k\rho}y(\rho)$ are strongly measurable on $(0,\infty)$. Moreover,

$$\|e^{-k\rho}y(\rho)\| \leqslant 2\pi\sqrt{2}\,[g]\,e^{-\varepsilon\rho}\|x(\rho)\|,\quad \rho>0$$

and

$$\int_0^\infty \|e^{-k\rho}y(\rho)\|\,d\rho < \infty.$$

Thus the $\bar{B}$-integral

$$Y = \int_0^\infty e^{-k\rho}y(\rho)\,d\rho \in L_2(\Omega)$$

exists. Let us set

$$y_1(\tau, \omega) = (1 + \mathrm{k}\omega\mathrm{n})^{-1}y(\tau, \omega), \quad Y'(\omega) = (1 + \mathrm{k}\omega\mathrm{n})^{-1}Y(\omega),$$

and, for every $v \in L_2(\Omega)$,

$$v_1(\omega) = \overline{(1 + \mathrm{k}\omega\mathrm{n})^{-1}}\, v(\omega).$$

We have

$$(y_1(\tau), v) = (y(\tau), v_1), \quad \tau > 0,$$

which implies that $y_1(\cdot)$ is a strongly measurable function.

At the same time,

$$\|e^{-k\tau}y_1(\tau)\| \leqslant \frac{1}{j_-}\|e^{-k\tau}y(\tau)\|, \quad j_- = \inf_{\omega\in\Omega} |1 + \mathrm{k}\omega\mathrm{n}| > 0$$

for $\tau > 0$. This means that $e^{-k\tau}y_1(\tau)$ is B-summable. We set

$$Y_1 = \int_0^\infty e^{-k\tau}y_1(\tau)\,d\tau.$$

By (53), for $v \in L_2(\Omega)$ we have

$$(Y', v) = (Y, v_1) = \int_0^\infty e^{-k\tau}(y(\tau), v_1)\,d\tau = \int_0^\infty e^{-k\tau}(y_1(\tau), v)\,d\tau = (Y_1, v),$$

which implies that $Y' = Y_1$. Now

$$(\tilde{z}(\mathrm{k}), u) = \int_0^\infty e^{-k\tau}(y_1(\tau), u)\,d\tau - \left(\frac{|\omega\mathrm{n}|}{1 + \mathrm{k}\omega\mathrm{n}} z^0, u\right) = (Y', u) - \left(\frac{|\omega\mathrm{n}|}{1 + \mathrm{k}\omega\mathrm{n}} z^0, u\right).$$

Since $u \in L_2(\Omega)$ is arbitrary, it follows that $\tilde{z}(\mathrm{k}) = Y'(|\omega\mathrm{n}|/(1 + \mathrm{k}\omega\mathrm{n}))\, z^0$. We infer from this that

$$(1 + \mathrm{k}\omega\mathrm{n})\,\tilde{z}(\mathrm{k}, \omega) = (1 + \mathrm{k}\omega\mathrm{n})\,Y'(\omega) - |\omega\mathrm{n}|\,z^0(\omega) = Y(\omega) - |\omega\mathrm{n}|\,z^0(\omega)$$

almost everywhere on Ω. But for $v \in L_2(\Omega)$ we have

$$(Y, v) = \int_0^\infty e^{-k\tau}(y(\tau), v)\,d\tau = \int_0^\infty e^{-k\tau}(x(\tau), \hat{g}v)\,d\tau = (\tilde{x}(\mathrm{k}), \hat{g}v) = (\hat{g}\tilde{x}(\mathrm{k}), v),$$

so that $Y = \hat{g}\tilde{x}(\mathrm{k})$,

$$(1 + \mathrm{k}\omega\mathrm{n})\,\tilde{z}(\mathrm{k}) = \hat{g}\tilde{x}(\mathrm{k}) - |\omega\mathrm{n}|\,z^0 = \hat{g}\tilde{x}(\mathrm{k}) + (\omega\mathrm{n})\,z^0,$$

and the lemma is proved.

Lemma 30. *Let* $\epsilon \in (-1, 1)$, $x(\cdot) \in \mathcal{B}_\epsilon$, *and let* $x(\cdot)$ *be a solution of Problem* I. *Then the limit* $x(0) = \lim_{\tau=0+0} x(\tau)$ *exists, and the Laplace image* $\widetilde{x}(\mathrm{k})$ *of the function* $x(\tau)$ *satisfies the equation*

$$(1 + \mathrm{k}\omega\mathbf{n})\,\widetilde{x}\,(\mathrm{k}) = \hat{g}\widetilde{x}\,(\mathrm{k}) + (\omega\mathbf{n})\,x\,(0) \tag{102}$$

for $\mathrm{k} \in Z_0$, $\mathrm{Re}\,\mathrm{k} > \epsilon$.

Proof. The existence of the limit $x(0)$ was established in Theorem 27. In Problem I the absolute term $f(\cdot)$ of equation (52) satisfies inclusion (93), which implies that the Laplace image $\widetilde{f}\,(\mathrm{k}) = \int_0^\infty e^{-\mathrm{k}\tau} f(\tau)\,d\tau$ is defined and depends analytically on k at least in the halfplane $\{\mathrm{k} | \mathrm{Re}\,\mathrm{k} > -1\}$.

Let $u \in L_2(\Omega)$. Then (compare Lemma 1)

$$(\widetilde{f}\,(\mathrm{k}),\, u) = \int_0^\infty e^{-\mathrm{k}\tau}\,(f\,(\tau),\, u)\,d\tau,$$

where

$$(f\,(\tau),\, u) = \int_0^1\int_0^{2\pi} B\,(\psi_\sigma(\mu,\,\varphi))\,e^{-\tau/\mu}\,\overline{u\,(\psi_\sigma\,(\mu,\,\varphi))}\,d\mu d\varphi.$$

As is easy to verify, the change in the order of integration yields the equation

$$\begin{aligned}(\widetilde{f}\,(\mathrm{k}),\, u) &= \int_0^1\int_0^{2\pi} B\,(\psi_\sigma\,(\mu,\,\varphi))\,\overline{u\,(\psi_\sigma\,(\mu,\,\varphi))}\,d\mu\,d\varphi \int_0^\infty e^{-\tau\left(\mathrm{k}+\frac{1}{\mu}\right)}d\tau \\ &= \int_\Omega \frac{(\omega\mathbf{n})\,B\,(\omega)}{1+\mathrm{k}\omega\mathbf{n}}\,\overline{u\,(\omega)}\,d\omega \equiv \left(\frac{\omega\mathbf{n}}{1+\mathrm{k}\omega\mathbf{n}}\,B,\, u\right).\end{aligned}$$

Recalling that u is an arbitrary element of $L_2(\Omega)$, we find that

$$\widetilde{f}\,(\mathrm{k},\,\omega) = \frac{\omega\mathbf{n}}{1+\mathrm{k}\omega\mathbf{n}}\,B\,(\omega)$$

almost everywhere on Ω.

We now set $z(\cdot) = \hat{A}x(\cdot)$, so that

$$x\,(\tau) = z\,(\tau) + f\,(\tau), \quad \tau > 0.$$

This clearly implies that for $\mathrm{k} \in Z_0$, $\mathrm{Re}\,\mathrm{k} > \epsilon$,

$$\widetilde{x}\,(\mathrm{k}) = \widetilde{z}\,(\mathrm{k}) + \widetilde{f}\,(\mathrm{k}) = \widetilde{z}\,(\mathrm{k}) + \frac{\omega\mathbf{n}}{1+\mathrm{k}\omega\mathbf{n}}\,B.$$

This means that (compare Lemma 29)

$$\begin{aligned}(1 + \mathrm{k}\omega\mathbf{n})\,\widetilde{x}\,(\mathrm{k},\,\omega) &= (1 + \mathrm{k}\omega\mathbf{n})\,\widetilde{z}\,(\mathrm{k},\,\omega) + (\omega\mathbf{n})\,B\,(\omega) \\ &= \hat{g}\,\widetilde{x}\,(\mathrm{k},\,\omega) + (\omega\mathbf{n})\,(z^0\,(\omega) + B\,(\omega)).\end{aligned}$$

almost everywhere on Ω. (According to our notation, $\hat{g}\tilde{x}(k, \omega)$ is the numerical value of the function $\hat{g}\tilde{x}(k)$ at the point $\omega \in \Omega$, i.e. $\hat{g}\tilde{x}(k, \omega) \equiv x^*_\omega(\hat{g}\tilde{x}(k))$.) Applying formula (100), we find that

$$(1 + k\omega\mathbf{n})\,\tilde{x}(k, \omega) = \hat{g}\tilde{x}(k, \omega) + (\omega\mathbf{n})\,x(0, \omega),$$
$$k \in Z_0, \quad \operatorname{Re} k > \varepsilon,$$

almost everywhere on Ω, Q.E.D.

We note now that equation (102) is simply the inhomogeneous characteristic equation (39) with an absolute term of special form,

$$\alpha(k, \omega) = (\omega\mathbf{n})\,x(0, \omega), \quad k \in Z_0, \ \omega \in \Omega.$$

On the basis of Theorem 16 we can say that for every $k \in Z_0 \setminus \mathfrak{R}$ the space $L_2(\Omega)$ contains exactly one solution $\psi(k)$ of equation (102):

$$(1 + k\omega\mathbf{n})\,\psi(k) = \hat{g}\psi(k) + (\omega\mathbf{n})\,x(0), \quad k \in Z_0 \setminus \mathfrak{R}.$$

Here $\psi(k)$ is a holomorphic function of the variable k in the domain $Z_0 \setminus \mathfrak{R}$. Since (102) is uniquely solvable in $Z_0 \setminus \mathfrak{R}$, we have $\psi(k) = \tilde{x}(k)$ for $\operatorname{Re} k > \epsilon$, $k \in Z_0 \setminus \mathfrak{R}$, which means that $\psi(k)$ is an analytic extension of $\tilde{x}(k)$ from the halfplane $\{k \mid \operatorname{Re} k > \epsilon\}$ onto the open set $Z_0 \setminus (\mathfrak{R} \cap (-1, \epsilon])$. We denote this analytic extension by $\tilde{x}(k)$.

Thus, if $x(\cdot)$ is a solution of Problem I which belongs to the class $\mathcal{B}_\epsilon$, $\epsilon \in (-1, 1)$, then the Laplace image $\tilde{x}(k)$, of the function $x(\tau)$ is an analytic function of k in the domain $Z_0 \setminus (\mathfrak{R} \cap (-1, \epsilon])$ and satisfies equation (102) at every point of this domain.

The behavior of $\tilde{x}(k)$ in the neighborhoods of the points of the set $\mathfrak{R} \cap (-1, \epsilon]$ can also be described with the aid of Theorem 16. Let $\lambda \in \mathfrak{R} \cap (-1, \epsilon]$. This means that some neighborhood S of the point λ lies entirely in $[Z_0 \setminus (\mathfrak{R} \cap (-1, \epsilon])] \cup \{\lambda\}$, and there exists a function $u(k)$ holomorphic in S and with values in $L_2(\Omega)$ such that for all $k \in S \setminus \{\lambda\}$

$$\tilde{x}(k) = -\frac{\operatorname{sgn}\lambda}{k - \lambda} \sum_{p=1}^{p_\lambda} (x(0), (\omega\mathbf{n})\,\psi_{\lambda p})\,\psi_{\lambda p} + u(k), \tag{103}$$

if $\lambda \neq 0$, and

$$\tilde{x}(k) = -\frac{3}{4\pi}\,\frac{1 - g_1}{k^2}\,(x(0), \omega\mathbf{n}) + \frac{3}{4\pi k}\,[(x(0), \omega\mathbf{n})\,\omega\mathbf{n} + (x(0), (\omega\mathbf{n})^2)] + u(k), \tag{104}$$

if $\lambda = 0$.

Further on we shall also require an estimate for $\|\tilde{x}(k)\|$ for large values of $\operatorname{Im} k$. To obtain this estimate we rewrite (102) as (compare (16))

$$\tilde{x}(\mathrm{k}) = \hat{U}(\mathrm{k})\tilde{x}(\mathrm{k}) + \frac{\omega\mathbf{n}}{1+\mathrm{k}\omega\mathbf{n}}x(0),\ \mathrm{k}\in Z_0\setminus\mathfrak{N}. \tag{105}$$

Let $\mathrm{k}_1 = \mathrm{Re}\,\mathrm{k}$, $\mathrm{k}_2 = \mathrm{Im}\,\mathrm{k} \neq 0$, $v \in L_2(\Omega)$, $v' = \hat{U}(\mathrm{k})\,v$. Then (compare Lemma 1)

$$\|v'\|^2 = \int_\Omega\left|\int_\Omega \frac{g(\omega\omega')}{1+\mathrm{k}\,\omega\mathbf{n}}v(\omega')\,d\omega'\right|^2 d\omega \leqslant 4\pi^2[g]^2\|v\|^2\int_{-1}^{1}\frac{d\mu}{|1+\mathrm{k}\mu|^2}.$$

This means that

$$\|\hat{U}(\mathrm{k})\|^2 \leqslant 4\pi^2[g]^2\int_{-1}^{1}\frac{d\mu}{|1+\mathrm{k}\mu|^2} < \frac{4\pi^3[g]^2}{|\mathrm{k}_2|}$$

Hence for $|\mathrm{k}_2| > 4\pi^3[g]^2$ solution (105) can be expressed by means of a Neumann series,

$$\tilde{x}(\mathrm{k}) = \sum_{\nu=0}^{\infty}\hat{U}^{\nu}(\mathrm{k})\left[\frac{\omega\mathbf{n}}{1+\mathrm{k}\omega\mathbf{n}}x(0)\right],$$

and hence

$$\|\tilde{x}(\mathrm{k})\| \leqslant \frac{\left\|\frac{\omega\mathbf{n}}{1+\mathrm{k}\omega\mathbf{n}}x(0)\right\|}{1-\|\hat{U}(\mathrm{k})\|} \leqslant \frac{\left\|\frac{\omega\mathbf{n}}{1+\mathrm{k}\omega\mathbf{n}}x(0)\right\|}{1-2\sqrt{\pi^3}[g]\,|\mathrm{k}_2|^{-1/2}}.$$

But for $\mathrm{k}_2 \neq 0$ we have $|\mu(1+\mathrm{k}\mu)^{-1}| \leq |\mathrm{k}_2|^{-1}$, $\mu \in [-1, 1]$, so that

$$\left\|\frac{\omega\mathbf{n}}{1+\mathrm{k}\omega\mathbf{n}}x(0)\right\| \leqslant \frac{1}{|\mathrm{k}_2|}\|x(0)\|.$$

Thus, if $x(\cdot)$ is a solution of Problem I, if $x(\cdot) \in \overline{\mathfrak{B}}$, and if $\tilde{x}(\mathrm{k})$ is the Laplace image of $x(\tau)$, then for $|\mathrm{Im}\,\mathrm{k}| > 16\pi^3[g]^2$ we have

$$\|\tilde{x}(\mathrm{k})\| \leqslant \frac{2\|x(0)\|}{|\mathrm{Im}\,\mathrm{k}|}. \tag{106}$$

We are now in a position to construct the asymptotic expansions of the solutions of Problem I.

Theorem 28. *Let* $\epsilon \in (-1, 1)$, *and let* $x(\cdot)$ *be a solution of Problem* I, $x(\cdot) \in \mathfrak{B}_\epsilon$. *We choose a* $\lambda^* \in (-1, \min\{0, \epsilon\})$, $\lambda^* \,\overline{\in}\, \mathfrak{N}$. *Then, for all* $\tau > 0$, a) *for* $g_0 < 1$ *and for* $g_0 = 1$, $\epsilon < 0$ *we have*

$$x(\tau) = -\sum_{\lambda\in\mathfrak{N}\cap(\lambda^*,\,\epsilon]}(\operatorname{sgn}\lambda)\,e^{\lambda\tau}\sum_{p=1}^{p_\lambda}(x(0), (\omega\mathbf{n})\,\psi_{\lambda p})\,\psi_{\lambda p} + e^{\lambda^*\tau}r_{\lambda^*}(\tau);$$

b) *for* $g_0 = 1$, $\epsilon \geq 0$,

$$x(\tau) = -\frac{3}{4\pi}(1-g_1)(x(0), \omega\mathbf{n})\,\tau + \frac{3}{4\pi}[(x(0), \omega\mathbf{n})\,\omega\mathbf{n} + (x(0), (\omega\mathbf{n})^2)] -$$

$$- \sum_{\substack{\lambda \in \mathfrak{N} \cap (\lambda^*, \varepsilon] \\ \lambda \neq 0}} (\operatorname{sgn} \lambda) e^{\lambda\tau} \sum_{p=1}^{p_\lambda} (x(0), (\omega n) \psi_{\lambda p}) \psi_{\lambda p} + e^{\lambda^* \tau} r_{\lambda^*}(\tau).$$

In both cases $\sup_{\tau>0} \|r_{\lambda^*}(\tau)\| < \infty$.

Proof. Let $\gamma \in (-1, 1)$. Let us define the contour $\mathfrak{L}(\gamma)$ on the complex k-plane,

$$\mathscr{L}(\gamma) = \{k \mid \operatorname{Re} k = \gamma, \ |\operatorname{Im} k| \leqslant 1 - \gamma\} \cup \{k \mid \operatorname{Re} k \leqslant \gamma, \ |\operatorname{Im} k| = 1 - \operatorname{Re} k\}.$$

Next, we establish the direction on $\mathfrak{L}(\gamma)$ which corresponds to increasing Im k. We denote the portion of $\mathfrak{L}(\gamma)$ lying between the straight lines $\operatorname{Im} k = \pm h$ by $\mathfrak{L}_h(\gamma)$,

$$\mathscr{L}_h(\gamma) = \{k \mid k \in \mathscr{L}(\gamma), \ |\operatorname{Im} k| \leqslant h\}.$$

It is clear that $\mathfrak{L}(\gamma) \subset Z_0$, $\gamma \in (-1, 1)$.

Let $u \in L_2(\Omega)$, and let $\tilde{x}(k)$ be the Laplace image of the function $x(\tau)$. The numerical function $(\tilde{x}(k), u)$ is holomorphic in the domain $Z_0 \setminus (\mathfrak{N} \cap (-1, \epsilon])$, where for $|\operatorname{Im} k| > 16\pi^3[g]^2$ we have

$$|(\tilde{x}(k), u)| \leqslant \frac{2\|u\| \, \|x(0)\|}{|\operatorname{Im} k|}$$

(compare (106)). This implies the absolute convergence of the integral

$$\int_{\mathscr{L}(\gamma)} e^{k\tau} (\tilde{x}(k), u) \, dk, \ \tau > 0, \ \gamma \in (-1, 1), \ \gamma \overline{\in} \mathfrak{N} \cap (-1, \varepsilon]. \qquad (107)$$

In fact, if either $\gamma \in (-1, 1) \setminus \mathfrak{N}$ or $\gamma > \epsilon$, then $(\tilde{x}(k), u)$ is regular on $\mathfrak{L}(\gamma)$; for $\tau > 0$ we have

$$\int_{\substack{\mathscr{L}(\gamma) \\ \operatorname{Re} k < -(1+16\pi^3[g]^2)}} |e^{k\tau} (\tilde{x}(k), u)| \, |dk| \leqslant 2 \int_{-\infty}^{-(1+16\pi^3[g]^2)} e^{\xi\tau} \cdot \frac{2\|u\| \, \|x(0)\|}{1 - \xi} \sqrt{2} \, d\xi < \infty.$$

We choose the numbers τ and γ in such a way as to ensure fulfillment of the conditions imposed on them in (107). For $h > 2 + 16\pi^3[g]^2$ we denote by $d^{\pm}(h)$ the straight-line segment connecting the point $1 - h \pm ih$ with the point $\gamma \pm ih$ and oriented in the direction of increasing Re k. Let

$$\varepsilon_h = \int_{d^+(h)} e^{k\tau} (\tilde{x}(k), u) \, dk - \int_{d^-(h)} e^{k\tau} (\tilde{x}(k), u) \, dk.$$

Then

$$|\varepsilon_h| \leqslant 2 \int_{1-h}^{\gamma} e^{\xi\tau} \frac{2\|u\| \, \|x(0)\|}{h} d\xi \leqslant \frac{4\|u\| \, \|x(0)\|}{h\tau} e^{\gamma\tau},$$

which implies that $\lim_{h=\infty} |\epsilon_h| = 0$.

It is clear that

$$\int_{\gamma-ih}^{\gamma+ih} e^{k\tau}(\tilde{x}(k), u)\,dk = \int_{\mathscr{L}_h(\gamma)} e^{k\tau}(\tilde{x}(k), u)\,dk + \varepsilon_h.$$

The existence of integral (107) implies that the right side of this relation has a limit as $h \to \infty$. This means that the integral on the left converges for $h = \infty$, and that

$$\int_{\gamma-i\infty}^{\gamma+i\infty} e^{k\tau}(\tilde{x}(k), u)\,dk = \int_{\mathscr{L}(\gamma)} e^{k\tau}(\tilde{x}(k), u)\,dk, \quad \tau > 0, \ \gamma \in (-1, 1), \ \gamma \overline{\in} \mathfrak{N} \cap (-1, \varepsilon]. \tag{108}$$

We note that the conditions of Theorems 28 and 27 coincide. Hence (compare Theorem 27 [ii]), $\sup_{\tau>0} e^{-\epsilon\tau}\|x(\tau)\| < \infty$. But Lemma 28 now implies that for $\gamma \in (\epsilon, 1)$ we have

$$e^{-\gamma\tau}(x(\tau), u) = \lim_{n=\infty} \frac{1}{2\pi i} \int_{\gamma-ih_n}^{\gamma+ih_n} e^{(k-\gamma)\tau}(\tilde{x}(k), u)\,dk,$$

for almost all $\tau \in (0, \infty)$, where $\{h_n\}$ is a specially chosen sequence of positive numbers, $h_n \to \infty$ as $n \to \infty$. Comparing this with formula (108) we find that

$$(x(\tau), u) = \frac{1}{2\pi i} \int_{\mathscr{L}(\gamma)} e^{k\tau}(\tilde{x}(k), u)\,dk, \quad \tau > 0, \ \gamma \in (\varepsilon, 1). \tag{109}$$

Let $\lambda^* \in (-1, \min\{0, \epsilon\})$, $\lambda^* \overline{\in} \mathfrak{N}$. Equation (109) can clearly be transformed into

$$(x(\tau), u) = \sum_{\lambda \in \mathfrak{N} \cap (\lambda^*, \varepsilon]} \operatorname*{res}_{k=\lambda} e^{k\tau}(\tilde{x}(k), u) + \frac{1}{2\pi i} \int_{\mathscr{L}(\lambda^*)} e^{k\tau}(\tilde{x}(k), u)\,dk, \quad \tau > 0. \tag{110}$$

If $\lambda \in \mathfrak{N} \cap (\lambda^*, \epsilon]$ and $\lambda \neq 0$, then, by (103),

$$\operatorname*{res}_{k=\lambda} e^{k\tau}(\tilde{x}(k), u) = -(\operatorname{sgn}\lambda)\, e^{\lambda\tau} \sum_{p=1}^{p_\lambda} (x(0), (\omega\mathbf{n})\,\psi_{\lambda p})(\psi_{\lambda p}, u).$$

But if $\lambda = 0$, then (compare (104))

$$\operatorname*{res}_{k=\lambda} e^{k\tau}(\tilde{x}(k), u) = -\frac{3}{4\pi}(1 - g_1)(x(0), \omega\mathbf{n})\,\tau\,(1, u)$$
$$+ \frac{3}{4\pi}((x(0), \omega\mathbf{n})\,\omega\mathbf{n} + (x(0), (\omega\mathbf{n})^2), u).$$

We now define the abstract function $r_{\lambda^*}(\cdot)$ by the following relations. If $g_0 < 1$ or $g_0 = 1$, $\epsilon < 0$, then let

$$e^{\lambda^*\tau} r_{\lambda^*}(\tau) = x(\tau) + \sum_{\lambda \in \mathfrak{N} \cap (\lambda^*, \varepsilon]} (\operatorname{sgn}\lambda)\, e^{\lambda\tau} \sum_{p=1}^{p_\lambda} (x(0), (\omega\mathbf{n})\,\psi_{\lambda p})\,\psi_{\lambda p}, \quad \tau > 0.$$

On the other hand, if $g_0 = 1$ and $\epsilon \geq 0$, then we set

$$e^{\lambda^*\tau} r_{\lambda^*}(\tau) = x(\tau) + \frac{3}{4\pi}(1-g_1)(x(0), \omega\mathbf{n})\tau - \frac{3}{4\pi}[(x(0), \omega\mathbf{n})\omega\mathbf{n} + (x(0), (\omega\mathbf{n})^2)] + \sum_{\substack{\lambda\in\mathfrak{N}\cap(\lambda^*, \varepsilon]\\ \lambda\neq 0}} (\operatorname{sgn}\lambda) e^{\lambda\tau} \sum_{p=1}^{p_\lambda} (x(0), (\omega\mathbf{n})\psi_{\lambda p})\psi_{\lambda p}.$$

It is clear that $r_{\lambda^*}(\cdot) \in \mathfrak{K}$. To prove the theorem we need merely establish the boundedness of $\|r_{\lambda^*}(\tau)\|$ on the halfline $\tau > 0$.

From the definition of $r_{\lambda^*}(\cdot)$ we see that

$$e^{\lambda^*\tau}\|r_{\lambda^*}(\tau)\| \leqslant \text{const}\cdot e^{\varepsilon\tau}, \quad \tau > 0,$$

which means that $\sup_{\tau\in(0,1]}\|r_{\lambda^*}(\tau)\| = a_1 < \infty$. On the other hand, (110) implies that for all $\tau > 0$

$$e^{\lambda^*\tau}(r_{\lambda^*}(\tau), u) = \frac{1}{2\pi i}\int_{\mathscr{L}(\lambda^*)} e^{k\tau}(\tilde{x}(k), u)\,dk.$$

By the choice of λ^*, the image $\tilde{x}(k)$ is continuous on the contour $\mathfrak{L}(\lambda^*)$. Together with estimate (106) this implies that

$$\sup_{k\in\mathscr{L}(\lambda^*)}\|\tilde{x}(k)\| = a_2 < \infty.$$

Thus for $\tau > 0$ we have

$$e^{\lambda^*\tau}|(r_{\lambda^*}(\tau), u)| \leqslant \frac{a_2\|u\|}{2\pi}\int_{\mathscr{L}(\lambda^*)} e^{\tau\operatorname{Re}k}|dk| = \frac{a_2\|u\|}{2\pi}\left\{2e^{\lambda^*\tau}(1-\lambda^*) + 2\sqrt{2}\int_{-\infty}^{\lambda^*} e^{\tau\xi}d\xi\right\}$$

and

$$|(r_{\lambda^*}(\tau), u)| \leqslant \frac{a_2\|u\|}{\pi}\left(1-\lambda^* + \frac{\sqrt{2}}{\tau}\right) \leqslant \frac{a_2\|u\|}{\pi}(1-\lambda^* + \sqrt{2}), \quad \tau \geqslant 1.$$

Since u is an arbitrary element of $L_2(\Omega)$, the latter inequality means that the set of vectors $\{r_*(\tau) | \tau \geq 1\}$ is weakly bounded in $L_2(\Omega)$. This implies the boundedness of this set in norm: $\sup_{\tau\geq 1}\|r_{\lambda^*}(\tau)\| < \infty$. The theorem is proved.

Theorem 29. *Let $\epsilon \in (-1, 1)$, and let $x(\cdot)$ be a solution of Problem* I, *$x(\cdot) \in \mathfrak{B}_\epsilon$. Then*

a) *For $g_0 < 1$ and for $g_0 = 1$, $\epsilon < 0$,*

$$x(\cdot) = -\sum_{\lambda\in\mathfrak{N}\cap(0,\varepsilon]}\sum_{p=1}^{p_\lambda}(x(0), (\omega\mathbf{n})\psi_{\lambda p})y_{\lambda p}(\cdot) + s_I(\cdot);$$

b) *For* $g_0 = 1, \epsilon \geq 0,$

$$x(\cdot) = -\sum_{\lambda \in \mathfrak{N} \cap (0, \varepsilon]} \sum_{p=1}^{p_\lambda} (x(0), (\omega n)\psi_{\lambda p}) y_{\lambda p}(\cdot) - \frac{3}{4\pi}(1-g_1)(x(0), \omega n) y_0(\cdot) + s_l(\cdot).$$

(*The functions* $y_{\lambda_p}(\cdot)$, $y_0(\cdot)$ *and* $s_l(\cdot)$ *were constructed in* §§9 *and* 10.)

Proof. We choose λ^* as in the preceding theorem and set

$$h_1(\tau) = x(\tau) + \sum_{\lambda \in \mathfrak{N} \cap (0, \varepsilon]} e^{\lambda\tau} \sum_{p=1}^{p_\lambda} (x(0), (\omega n)\psi_{\lambda p})\psi_{\lambda p}, \quad \tau > 0,$$

in case a), and

$$h_1(\tau) = x(\tau) + \frac{3}{4\pi}(1-g_1)(x(0), \omega n)\tau + \sum_{\lambda \in \mathfrak{N} \cap (0, \varepsilon]} e^{\lambda\tau} \sum_{p=1}^{p_\lambda} (x(0), (\omega n)\psi_{\lambda p})\psi_{\lambda p}$$

in case b).

Making use of the formulas of Theorem 28, we find that $\sup_{\tau > 0} \|h_1(\tau)\| < \infty$. We can now write (compare Theorem 23 and Lemma 24)

$$x(\tau) = -\sum_{\lambda \in \mathfrak{N} \cap (0, \varepsilon]} \sum_{p=1}^{p_\lambda} (x(0), (\omega n)\psi_{\lambda p}) y_{\lambda p}(\tau) + h_1(\tau) + h_2(\tau),$$

$$h_2(\tau) = \sum_{\lambda \in \mathfrak{N} \cap (0, \varepsilon]} \sum_{p=1}^{p_\lambda} (x(0), (\omega n)\psi_{\lambda p})(y_{\lambda p}(\tau) - x_{\lambda p}(\tau))$$

in case a), and

$$x(\tau) = -\sum_{\lambda \in \mathfrak{N} \cap (0, \varepsilon]} \sum_{p=1}^{p_\lambda} (x(0), (\omega n)\psi_{\lambda p}) y_{\lambda p}(\tau)$$

$$-\frac{3}{4\pi}(1-g_1)(x(0), \omega n) y_0(\tau) + h_1(\tau) + h_2(\tau),$$

$$h_2(\tau) = \sum_{\lambda \in \mathfrak{N} \cap (0, \varepsilon]} \sum_{p=1}^{p_\lambda} (x(0), (\omega n)\psi_{\lambda p})(y_{\lambda p}(\tau) - x_{\lambda p}(\tau))$$

$$+\frac{3}{4\pi}(1-g_1)(x(0), \omega n)(y_0(\tau) - x^{(0)}(\tau))$$

in case b).

Theorem 23[ii] and Lemma 24[ii] imply that $\sup_{\tau > 0} \|h_2(\tau)\| < \infty$. Since every function $y_{\lambda_p}(\cdot)$ (and for $g_0 = 1$ the function $y_0(\cdot)$ as well) is a solution of the homogeneous Milne problem, it follows that the sum $h_1(\cdot) + h_2(\cdot)$ is a solution of Problem I. The inequality $\sup_{\tau > 0} \|h_1(\tau) + h_2(\tau)\| < \infty$ established above means that this solution belongs to the class $\mathfrak{T}_{\lambda_0}$ (see §9), and thus necessarily coincides with $s_l(\cdot)$ (compare (97)). The theorem is proved.

Theorem 30. *Let* $x(\cdot) \in \overline{\mathfrak{B}}$. *The function* $x(\cdot)$ *is a solution of the homogeneous Milne problem if and only if* $x(\cdot)$ *is a linear combination of a finite number of functions from the system* $\mathfrak{X}$.

Proof. If $x(\cdot)$ is a linear combination of elements of $\mathfrak{X}$, then $x(\cdot)$ is clearly a solution of the homogeneous Milne problem.

Conversely, let $x(\cdot)$ be a solution of the homogeneous Milne problem, or, which is the same thing, of homogeneous Problem I, $x(\cdot) \in \mathfrak{B}$. It is clear that the solution of homogeneous Problem I from the class $\mathfrak{T}_{\lambda_0}$ is identically zero: $s_I(\tau) \equiv \vartheta$, $\tau > 0$ (compare §10). $x(\cdot) \in \mathfrak{B}_\epsilon$ for some $\epsilon \in (-1, 1)$. Theorem 30 now follows from the formulas of Theorem 29 and from the finiteness of the set $\mathfrak{R} \cap (0, \epsilon]$ (Theorem 5).

These results enable us to strengthen Lemma 25 (and Theorem 26) somewhat. Various (close, but not equivalent) formulations are possible here. The following is a typical example.

We note that for $g_0 < 1$ we have $\mathfrak{T}_{\lambda_0} \subset \mathfrak{B}_{\lambda_0}$, but that the reverse inclusion is not valid. Nevertheless, Lemma 25 remains valid if we replace $\mathfrak{T}_{\lambda_0}$ by $\mathfrak{B}_{\lambda_0}$ in its formulation. In order to see this we first assume that $x(\cdot)$ is a solution of the homogeneous Milne problem, $x(\cdot) \in \mathfrak{B}_{\lambda_0}$. If $g_0 = 1$, then $\mathfrak{B}_{\lambda_0} = \mathfrak{B}_0 \subset \mathfrak{T}_{\lambda_0}$, and no proof is required. But if $g_0 < 1$, then, by formula a) of Theorem 28,

$$x(\tau) = -e^{\lambda_0\tau}(x(0), (\omega\mathbf{n})\psi_{\lambda_0,1})\psi_{\lambda_0,1} + e^{\lambda^*\tau} r_{\lambda^*}(\tau),$$
$$\lambda^* \in (-\lambda_0, 0), \ \sup_{\tau>0} \| r_{\lambda^*}(\tau) \| = a < \infty.$$

This implies that

$$e^{-\lambda_0\tau}\| x(\tau) \| \geqslant |(x(0), (\omega\mathbf{n})\psi_{\lambda_0,1})| \, \| \psi_{\lambda_0,1} \| - a e^{-(\lambda_0-\lambda^*)\tau}.$$

But, by hypothesis, the function on the left is summable. This means that

$$(x(0), (\omega\mathbf{n})\psi_{\lambda_0,1}) = 0, \text{ and } \| x(\tau) \| \leqslant \| r_{\lambda^*}(\tau) \| \leqslant a, \ \tau > 0.$$

Thus $x(\cdot) \in \mathfrak{T}_{\lambda_0}$, which implies that $x(\tau) \equiv \vartheta$ (compare Lemma 25, the old formulation).

Theorems 28–30 yield still other interesting results. We shall consider these presently.

Lemma 31. *Let $\epsilon \in (-1, 1)$, and let $x(\cdot)$ be a solution of Problem* I, $x(\cdot) \in \mathfrak{B}_\epsilon$. *Then*

[i] *If $\lambda \in \mathfrak{R}$, $\lambda \neq 0$, $\lambda \geq \epsilon$, then for all $p = 1, \cdots, p_\lambda$ we have*

$$(x(0), (\omega\mathbf{n})\psi_{\lambda p}) = 0.$$

[ii] *If $g_0 = 1$, $\epsilon \leq 0$, then*

$$(x(0), \omega\mathbf{n}) = (x(0), (\omega\mathbf{n})^2) = 0.$$

[iii] *If $\lambda \in \mathfrak{R}$, $\lambda > 0$, then*

$$(s_I(0), (\omega\mathbf{n})\psi_{\lambda p}) = 0$$

for all $p = 1, \cdots, p_\lambda$. *If, in addition,* $g_0 = 1$, *then* $(s_l(0), \omega\mathbf{n}) = 0$.

Proof. To simplify our expressions we set

$$a_{\lambda p} = (x(0), (\omega\mathbf{n})\psi_{\lambda p}).$$

Let the conditions of [i] be fulfilled. Then, obviously, $x(\cdot) \in \mathfrak{B}_\lambda$. Theorem 28 implies that for $g_0 < 1$ and for $g_0 = 1$, $\lambda < 0$ we have

$$e^{-\lambda\tau}x(\tau) = -(\operatorname{sgn}\lambda)\sum_{p=1}^{p_\lambda} a_{\lambda p}\psi_{\lambda p}$$

$$-\sum_{\lambda' \in \mathfrak{N}\cap(\lambda^*,\lambda)} (\operatorname{sgn}\lambda')e^{-(\lambda-\lambda')\tau}\sum_{p=1}^{p_{\lambda'}} a_{\lambda' p}\psi_{\lambda' p} + e^{-(\lambda-\lambda^*)\tau}r_{\lambda^*}(\tau).$$

But if $g_0 = 1$, $\lambda > 0$, then

$$e^{-\lambda\tau}x(\tau) = -\sum_{p=1}^{p_\lambda} a_{\lambda p}\psi_{\lambda p} - \sum_{\substack{\lambda' \in \mathfrak{N}\cap(\lambda^*,\lambda)\\ \lambda' \neq 0}} (\operatorname{sgn}\lambda')e^{-(\lambda-\lambda')\tau}\sum_{p=1}^{p_{\lambda'}} a_{\lambda' p}\psi_{\lambda' p}$$

$$-\frac{3}{4\pi}(1-g_1)(x(0), \omega\mathbf{n})\tau e^{-\lambda\tau} + \frac{3}{4\pi}e^{-\lambda\tau}[(x(0), \omega\mathbf{n})\omega\mathbf{n} + (x(0), (\omega\mathbf{n})^2)]$$
$$+ e^{-(\lambda-\lambda^*)\tau}r_{\lambda^*}(\tau).$$

In both cases $\lambda^* \in (-1, \min\{0, \lambda\})$, $\lambda^* \overline{\in} \mathfrak{N}$ and $\sup_{\tau>0}\|r_{\lambda^*}(\tau)\| < \infty$. The above expressions imply that

$$\lim_{\tau=\infty} e^{-\lambda\tau}x(\tau) = -(\operatorname{sgn}\lambda)\sum_{p=1}^{p_\lambda} a_{\lambda p}\psi_{\lambda p}.$$

But the function $e^{-\lambda\tau}\|x(\tau)\|$ is summable. Hence $\sum_{p=1}^{p_\lambda} a_{\lambda p}\psi_{\lambda p} = \vartheta$, and, owing to the linear independence of the vectors $\psi_{\lambda p}$, $p = 1, \cdots, p_\lambda$,

$$a_{\lambda p} = 0,\ p = 1, 2, \ldots, p_\lambda.$$

Assertion [i] is proved.

Now let us assume that $g_0' = 1$, $\epsilon \le 0$. This means that $x(\cdot) \in \mathfrak{B}_0$, and, by part b) of Theorem 28,

$$x(\tau) = -\frac{3}{4\pi}(1-g_1)(x(0), \omega\mathbf{n})\tau + \frac{3}{4\pi}[(x(0), \omega\mathbf{n})\omega\mathbf{n} + (x(0), (\omega\mathbf{n})^2)]$$
$$+ e^{\lambda^*\tau}r_{\lambda^*}(\tau).$$

Here the number $\lambda^* \in (-1, 0)$ is chosen in such a way that the interval $[\lambda^*, 0)$ does not contain points from the set $\mathfrak{N}$; $\sup_{\tau>0}\|r_{\lambda^*}(\tau)\| < \infty$. Clearly,

$$\lim_{\tau=\infty}(\tau+1)^{-1}x(\tau) = -\frac{3}{4\pi}(1-g_1)(x(0), \omega\mathbf{n}).$$

On the other hand, the inclusion $x(\cdot) \in \mathfrak{B}_0$ implies the summability of the function $(\tau+1)^{-1}\|x(\tau)\|$, which implies that $(1-g_1)(x(0), \omega\mathbf{n}) = 0$. But $g_1 < g_0 = 1$

(compare (14)). This means that

$$(x(0),\ \omega\mathbf{n}) = 0.$$

Now

$$x(\tau) = \frac{3}{4\pi}(x(0),\ (\omega\mathbf{n})^2) + e^{\lambda^*\tau} r_{\lambda^*}(\tau),$$

so that $\lim_{\tau=\infty} x(\tau) = (3/4\pi)(x(0), (\omega\mathbf{n})^2)$. This limit relation and the summability of $\|x(\tau)\|$ imply that

$$(x(0),\ (\omega\mathbf{n})^2) = 0$$

as well.

Now let us prove [iii]. Let $\lambda \in \mathfrak{N}$, $\lambda > 0$. By (98) there exists an $\epsilon \in (-1, \lambda)$ such that $s_I(\cdot) \in \mathfrak{B}_\epsilon$. The equations $(s_I(\cdot),\ (\omega\mathbf{n})\psi_{\lambda_p}) = 0$, $p = 1, 2, \cdots, p_\lambda$ are now a special case of statement [i], which we have already proved.

Let us assume that $g_0 = 1$. We choose $\epsilon \in (0, 1)$ and $\lambda^* \in (-1, 0)$ in such a way that the segment $[\lambda^*, \epsilon]$ contains only one point of the set $\mathfrak{N}$, i.e. the point 0. Since $s_I(\cdot) \in \mathfrak{B}_\epsilon$ (compare (98)), it follows by formula b) of Theorem 28 that

$$s_I(\tau) = -\frac{3}{4\pi}(1 - g_1)(s_I(0),\ \omega\mathbf{n})\tau + \frac{3}{4\pi}[(s_I(0),\ \omega\mathbf{n})\,\omega\mathbf{n} + (s_I(0),\ (\omega\mathbf{n})^2)] + e^{\lambda^*\tau} r_{\lambda^*}(\tau).$$

The equation $(s_I(0),\ \omega\mathbf{n}) = 0$ now follows from the boundedness of $\|s_I(\tau)\|$ (estimate (99)). The lemma is proved.

Lemma 31 enables us to refine the notation of the formulas of Theorems 28 and 29. In fact, in all of these formulas we can contract the sets of values of the parameter λ over which summation is carried out by removing from them the possible value $\lambda = \epsilon$ (by Lemma 31 the corresponding term is equal to zero). For example, summation in the formulas of Theorem 29 is carried out not over $\lambda \in \mathfrak{N} \cap (0, \epsilon]$, but rather over $\lambda \in \mathfrak{N} \cap (0, \epsilon)$.

It is interesting to compare the orthogonality relations of Lemma 31 with certain integrals of the solutions of Problem I. We are referring to the constancy of the radiation flux and to the linearity of the so-called K-integral [6] in the conservative case, as well as to the corresponding generalizations for the nonconservative case.

Lemma 32. *Let* $\epsilon \in (-1, 1)$ *and let* $x(\cdot) \in \mathfrak{B}_\epsilon$ *be a solution of Problem* I; *then*

[i] *If* $\lambda \in \mathfrak{N}$ *and* $\psi \in H_\lambda$, *then for* $\tau > 0$ *we have*

$$(x(\tau),\ (\omega\mathbf{n})\psi) = e^{\lambda\tau}(x(0),\ (\omega\mathbf{n})\psi).$$

[ii] *In the conservative case* ($g_0 = 1$), *for* $\tau > 0$ *we have*

$$(x(\tau),\ \omega\mathbf{n}) = (x(0),\ \omega\mathbf{n});$$
$$(x(\tau),\ (\omega\mathbf{n})^2) = -(1-g_1)(x(0),\ \omega\mathbf{n})\,\tau + (x(0),\ (\omega\mathbf{n})^2).$$

Proof. Let $\lambda \in \mathfrak{N}$, $\psi \in H_\lambda$. Then (compare Lemma 14)

$$(1 + \lambda\omega\mathbf{n})\,\psi = \hat{g}\psi.$$

For $\operatorname{Re} k > \epsilon$ the Laplace image $x(k)$ of the function $x(\cdot)$ satisfies equation (102). Let us take the scalar product of the right side of this equation and ψ:

$$(\tilde{x}(k),\ \psi) + k\,(\tilde{x}(k),\ (\omega\mathbf{n})\,\psi) = (\hat{g}\tilde{x}(k),\ \psi) + (x(0),\ (\omega\mathbf{n})\,\psi)$$
$$= (\tilde{x}(k),\ (1+\lambda\omega\mathbf{n})\,\psi) + (x(0),\ (\omega\mathbf{n})\,\psi),$$

from which we have

$$(\tilde{x}(k),\ (\omega\mathbf{n})\,\psi) = \frac{1}{k-\lambda}(x(0),\ (\omega\mathbf{n})\,\psi).$$

On the left we have the Laplace transform of the numerical function $(x(\tau), (\omega\mathbf{n})\psi)$ (compare (101′)). Statement [i] is now self-evident.

Let us assume that $g_0 = 1$. We take the scalar product of (102) and a function from $L_2(\Omega)$ which is identically equal to unity,

$$(\tilde{x}(k),\ 1) + k\,(\tilde{x}(k),\ \omega\mathbf{n}) = (\tilde{x}(k),\ \hat{g}1) + (x(0),\ \omega\mathbf{n}),$$

which means that

$$(\tilde{x}(k),\ \omega\mathbf{n}) = \frac{1}{k}(x(0),\ \omega\mathbf{n}),\ \operatorname{Re} k > \varepsilon,$$

and therefore

$$(x(\tau),\ \omega\mathbf{n}) = (x(0),\ \omega\mathbf{n}) \quad \text{for } \tau > 0$$

(constancy of the flux).

Taking the scalar product of (102) and $\omega\mathbf{n}$, we obtain

$$(\tilde{x}(k),\ \omega\mathbf{n}) + k\,(\tilde{x}(k),\ (\omega\mathbf{n})^2) = g_1(\tilde{x}(k),\ \omega\mathbf{n}) + (x(0),\ (\omega\mathbf{n})^2),$$

so that

$$(\tilde{x}(k),\ (\omega\mathbf{n})^2) = -\frac{1-g_1}{k}(\tilde{x}(k),\ \omega\mathbf{n}) + \frac{1}{k}(x(0),\ (\omega\mathbf{n})^2)$$
$$= -\frac{1-g_1}{k^2}(x(0),\ \omega\mathbf{n}) + \frac{1}{k}(x(0),\ (\omega\mathbf{n})^2),\ \operatorname{Re} k > \varepsilon.$$

This relation implies the second formula of statement [ii] (linearity of the K-integral).

Now let us compare Lemmas 32 and 31, assuming as before that $x(\cdot)$ is a solution of Problem I, $x(\cdot) \in \mathfrak{B}_\epsilon$, $\epsilon \in (-1, 1)$. We arrive at the following self-evident conclusions.

If $\lambda \in \mathfrak{N}$, $\lambda \neq 0$, $\lambda \geq \epsilon$, then

$$(x(\tau), (\omega \mathbf{n}) \psi_{\lambda p}) = 0, \ \tau > 0, \ p = 1, 2, \ldots, p_\lambda;$$

if $g_0 = 1$ and $\epsilon \leq 0$, then for all $\tau > 0$ we have

$$(x(\tau), \omega \mathbf{n}) = (x(\tau), (\omega \mathbf{n})^2) = 0.$$

Specifically, for $\lambda \in \mathfrak{N}$, $\lambda > 0$,

$$(s_I(\tau), (\omega \mathbf{n}) \psi_{\lambda p}) = 0, \ \tau > 0.$$

But if $g_0 = 1$, then

$$(s_I(\tau), \omega \mathbf{n}) = 0, \ \tau > 0.$$

In conclusion let us see what new information about the properties of the functions of the system $\mathfrak{X}$ can be extracted from the results obtained in this section.

Let $\lambda \in \mathfrak{N}$, $\lambda > 0$, $1 \leq p \leq p_\lambda$.

We recall that the system $\mathfrak{X}$ is linearly independent (Theorem 24) and that every function of this system is a solution of homogeneous Problem I for which $s_I(\tau) \equiv \vartheta$, $\tau > 0$. This and the formulas of Theorem 29 immediately imply that

$$(y_{\lambda p}(0), (\omega \mathbf{n}) \psi_{\lambda' p'}) = -\delta_{\lambda\lambda'} \delta_{pp'}, \ \lambda' \in \mathfrak{N} \cap (0, 1), \ 1 \leqslant p' \leqslant p_{\lambda'}.$$

If, in addition, $g_0 = 1$, then

$$(y_{\lambda p}(0), \omega \mathbf{n}) = 0;$$
$$(y_0(0), (\omega \mathbf{n}) \psi_{\lambda p}) = 0;$$
$$(y_0(0), \omega \mathbf{n}) = -\frac{4\pi}{3} \frac{1}{1 - g_1}.$$

From Lemma 32 and these relations we obtain

$$(y_{\lambda p}(\tau), (\omega \mathbf{n}) \psi_{\lambda' p'}) = -\delta_{\lambda\lambda'} \delta_{pp'} e^{\lambda\tau}, \ \tau > 0, \ \lambda' \in \mathfrak{N} \cap (0,1), \ 1 \leqslant p' \leqslant p_{\lambda'}.$$

For $g_0 = 1$ we have

$$(y_{\lambda p}(\tau), \omega \mathbf{n}) = 0, \ \tau > 0;$$
$$(y_{\lambda p}(\tau), (\omega \mathbf{n})^2) = (y_{\lambda p}(0), (\omega \mathbf{n})^2), \ \tau > 0;$$
$$(y_0(\tau), (\omega \mathbf{n}) \psi_{\lambda p}) = 0, \ \tau > 0;$$
$$(y_0(\tau), \omega \mathbf{n}) = -\frac{4\pi}{3} \frac{1}{1 - g_1}, \ \tau > 0.$$

We can now write out the following asymptotic expansions for the functions of the system $\mathfrak{X}$ (compare Theorem 28). If $\lambda \in \mathfrak{N}$, $\lambda > 0$, $1 \leq p \leq p_\lambda$, then for all $(\tau, \omega) \in (0, \infty) \times \Omega$

a) for $g_0 < 1$,

$$y_{\lambda p}(\tau, \omega) = e^{\lambda\tau}\psi_{\lambda p}(\omega) + \sum_{\lambda' \in \mathfrak{N} \cap (\lambda^*, 0)} e^{\lambda'\tau} \sum_{p'=1}^{p_{\lambda'}} (y_{\lambda p}(0), (\omega \mathbf{n}) \psi_{\lambda' p'}) \psi_{\lambda' p'}(\omega) + e^{\lambda^*\tau} r_{\lambda^*}(\tau, \omega); \tag{111}$$

b) for $g_0 = 1$.

$$y_{\lambda p}(\tau, \omega) = e^{\lambda\tau}\psi_{\lambda p}(\omega) + \frac{3}{4\pi}(y_{\lambda p}(0), (\omega \mathbf{n})^2) + \sum_{\lambda' \in \mathfrak{N} \cap (\lambda^*, 0)} e^{\lambda'\tau} \sum_{p'=1}^{p_{\lambda'}} (y_{\lambda p}(0), (\omega \mathbf{n}) \psi_{\lambda' p'}) \psi_{\lambda' p'}(\omega) + e^{\lambda^*\tau} r_{\lambda^*}(\tau, \omega); \tag{112}$$

$$y_0(\tau, \omega) = \tau - \frac{\omega \mathbf{n}}{1 - g_1} + \frac{3}{4\pi}(y_0(0), (\omega \mathbf{n})^2) + \sum_{\lambda' \in \mathfrak{N} \cap (\lambda^*, 0)} e^{\lambda'\tau} \sum_{p'=1}^{p_{\lambda'}} (y_0(0), (\omega \mathbf{n}) \psi_{\lambda' p'}) \psi_{\lambda' p'}(\omega) + e^{\lambda^*\tau} \overset{0}{r}_{\lambda^*}(\tau, \omega). \tag{113}$$

In these formulas $\lambda^* \in (-1, 0)$, $\lambda^* \bar{\in} \mathfrak{N}$, and

$$\sup_{\tau > 0} \| r_{\lambda^*}(\tau) \| < \infty, \quad \sup_{\tau > 0} \| \overset{0}{r}_{\lambda^*}(\tau) \| < \infty.$$

From formulas (111)–(113) we see that $r_{\lambda^*}(\tau, \omega)$ and $\overset{0}{r}_{\lambda^*}(\tau, \omega)$ are continuous functions of (τ, ω) on $(0, \infty) \times \Omega$ (compare Theorem 23[iii] and Lemma 24[iii]). Let us show that these functions are, in fact, bounded not only in norm, but also uniformly.

Let $k = \sup_{\tau > 0} \| r_{\lambda^*}(\tau) \|$. If $g_0 < 1$, then

$$k = \max \left\{ \sup_{\tau > 0} \| r_{\lambda^*}(\tau) \|, \ \sup_{\tau > 0} \| \overset{0}{r}_{\lambda^*}(\tau) \| \right\}, \quad \text{if} \quad g_0 = 1.$$

In any case $0 \le k < \infty$. For all $(\tau, \omega) \in (0, \infty) \times \Omega$ (compare Theorem 1) we have

$$| \overset{*}{x}_\omega (\hat{g} e^{\lambda^*\tau} r_{\lambda^*}(\tau)) | \leqslant \sqrt{2\pi}\, [g]\, k e^{\lambda^*\tau}.$$

From formulas (61) and (57)–(59) we now readily obtain

$$| \hat{A} e^{\lambda^*\tau} r_{\lambda^*}(\tau, \omega) | \leqslant \frac{\sqrt{2\pi} k [g]}{1 + \lambda^*} e^{\lambda^*\tau}, \quad (\tau, \omega) \in (0, \infty) \times \Omega.$$

A precisely similar estimate for $g_0 = 1$ also holds for $\hat{A} e^{\lambda^*\tau} \overset{0}{r}_{\lambda^*}(\tau, \omega)$.

Let us assume that $g_0 < 1$. Letting the operator $\hat{A}$ act on (111) and recalling that $\hat{A} y_{\lambda_p}(\cdot) = y_{\lambda_p}(\cdot)$, we obtain

$$\hat{A} e^{\lambda\tau}\psi_{\lambda p}(\omega) = e^{\lambda\tau}\psi_{\lambda p}(\omega) - \psi_{\lambda p}(\omega)\, \eta(\omega \mathbf{n})\, e^{-\frac{\tau}{\omega \mathbf{n}}}, \quad \lambda \in \mathfrak{N}, \ 1 \leqslant p \leqslant p_\lambda$$

(compare (76)). This yields the equation

$$r_{\lambda^*}(\tau, \omega) = -\eta(\omega\mathbf{n})\psi_{\lambda p}(\omega)e^{-\frac{\tau}{\omega\mathbf{n}}-\lambda^*\tau} + e^{-\lambda^*\tau}\hat{A}e^{\lambda^*\tau}r_{\lambda^*}(\tau, \omega)$$
$$-\sum_{\lambda' \in \mathfrak{N}\cap(\lambda^*,\,0)}\sum_{p'=1}^{p_{\lambda'}}(y_{\lambda p}(0),\ (\omega\mathbf{n})\psi_{\lambda' p'})\psi_{\lambda' p'}(\omega)\eta(\omega\mathbf{n})e^{-\frac{\tau}{\omega\mathbf{n}}-\lambda^*\tau},$$

which means that

$$|r_{\lambda^*}(\tau, \omega)| \leqslant \psi^+_{\lambda p} + \sum_{\lambda' \in \mathfrak{N}\cap(\lambda^*,\,0)}\sum_{p'=1}^{p_{\lambda'}}|(y_{\lambda p}(0),\ (\omega\mathbf{n})\psi_{\lambda' p'})|\,\psi^+_{\lambda' p'}$$
$$+\frac{\sqrt{2\pi}\,k[g]}{1+\lambda^*} = \text{const} < \infty,\quad (\tau, \omega) \in (0, \infty) \times \Omega.$$

For $g_0 = 1$ we find (compare Lemma 24, for example) in a similar way that

$$e^{\lambda^*\tau}r_{\lambda^*}(\tau, \omega) = -\psi_{\lambda p}(\omega)\eta(\omega\mathbf{n})e^{-\frac{\tau}{\omega\mathbf{n}}} - \frac{3}{4\pi}(y_{\lambda p}(0),\ (\omega\mathbf{n})^2)\eta(\omega\mathbf{n})e^{-\frac{\tau}{\omega\mathbf{n}}}$$
$$-\sum_{\lambda' \in \mathfrak{N}\cap(\lambda^*,0)}\sum_{p'=1}^{p_{\lambda'}}(y_{\lambda p}(0),\ (\omega\mathbf{n})\psi_{\lambda' p'})\psi_{\lambda' p'}(\omega)\eta(\omega\mathbf{n})e^{-\frac{\tau}{\omega\mathbf{n}}} + \hat{A}e^{\lambda^*\tau}r_{\lambda^*}(\tau, \omega);$$
$$e^{\lambda^*\tau}r^0_{\lambda^*}(\tau, \omega) = \left(\frac{\omega\mathbf{n}}{1-g_1} - \frac{3}{4\pi}(y_0(0),\ (\omega\mathbf{n})^2)\right)\eta(\omega\mathbf{n})e^{-\frac{\tau}{\omega\mathbf{n}}}$$
$$-\sum_{\lambda' \in \mathfrak{N}\cap(\lambda^*,0)}\sum_{p'=1}^{p_{\lambda'}}(y_0(0),\ (\omega\mathbf{n})\psi_{\lambda' p'})\psi_{\lambda' p'}(\omega)\eta(\omega\mathbf{n})e^{-\frac{\tau}{\omega\mathbf{n}}} + \hat{A}e^{\lambda^*\tau}r^0_{\lambda^*}(\tau, \omega).$$

Hence

$$|r_{\lambda^*}(\tau, \omega)| \leqslant \psi^+_{\lambda p} + \frac{3}{4\pi}|(y_{\lambda p}(0),\ (\omega\mathbf{n})^2)|$$
$$+\sum_{\lambda' \in \mathfrak{N}\cap(\lambda^*,0)}\sum_{p'=1}^{p_{\lambda'}}|(y_{\lambda p}(0),\ (\omega\mathbf{n})\psi_{\lambda' p'})|\,\psi^+_{\lambda' p'} + \frac{\sqrt{2\pi}k[g]}{1+\lambda^*} = \text{const} < \infty;$$
$$|r^0_{\lambda^*}(\tau, \omega)| \leqslant \frac{1}{1-g_1} + \frac{3}{4\pi}|(y_0(0),\ (\omega\mathbf{n})^2)|$$
$$+\sum_{\lambda' \in \mathfrak{N}\cap(\lambda^*,0)}\sum_{p'=1}^{p_{\lambda'}}|(y_0(0),\ (\omega\mathbf{n})\psi_{\lambda' p'})|\,\psi^+_{\lambda' p'} + \frac{\sqrt{2\pi}k[g]}{1+\lambda^*} = \text{const} < \infty;$$
$$(\tau, \omega) \in (0, \infty) \times \Omega.$$

Thus, in formulas (111)–(113),

$$\sup_{\tau>0}\|r_{\lambda^*}(\tau)\|_{C(\Omega)} < \infty\ (g_0 \leqslant 1);\quad \sup_{\tau>0}\|r^0_{\lambda^*}(\tau)\|_{C(\Omega)} < \infty\ (g_0 = 1).$$

Asymptotic expansions (111) and (113) of the solution $J(\cdot) \in \mathfrak{X}$ of fixed sign (compare Theorem 24) can be put into particularly simple form:[1)]

---

1) Similar expansions in the azimuthally homogeneous case were obtained by the present author [22] and in somewhat different form by Marčuk and Nikolaišvili [45].

a) *For* $g_0 < 1$ *and* $(\tau, \omega) \in (0, \infty) \times \Omega$,

$$J(\tau, \omega) = e^{\lambda_0 \tau} \psi_{\lambda_0, 1}(\omega) + (J(0), (\omega \mathbf{n}) \psi_{-\lambda_0, 1}) e^{-\lambda_0 \tau} \psi_{-\lambda_0, 1}(\omega) + e^{-\lambda_1 \tau} \rho_{\lambda_1}(\tau, \omega)$$

(cf. (40)).

b) *For* $g_0 = 1$ *and* $(\tau, \omega) \in (0, \infty) \times \Omega$,

$$J(\tau, \omega) = \tau - \frac{\omega \mathbf{n}}{1 - g_1} + \frac{3}{4\pi} (J(0), (\omega \mathbf{n})^2) + e^{-\lambda_1 \tau} \rho_{\lambda_1}(\tau, \omega).$$

In both cases $\sup_{\tau > 0} \| \rho_{\lambda_1}(\tau) \|_{C(\Omega)} < \infty$; λ_1 is the smallest number in the set $\mathfrak{N} \cap (\lambda_0, 1)$ if such numbers exist; otherwise λ_1 is an arbitrary number from the interval $(\lambda_0, 1)$.

§12. Some asymptotic properties of the solutions of the characteristic equation

It follows from the results of the preceding section that the solid-angle distribution of radiation in the homogeneous nonconservative Milne problem at large optical depths is of the form

$$J(\tau, \omega) = C e^{\lambda_0 \tau} \varphi_{\lambda_0}(\omega \mathbf{n}) + O(e^{-\lambda_0 \tau}).$$

Here C is a constant, $\phi_k(\mu)$ is the function constructed in Theorem 15, and $\mu \in [-1, 1]$, $k \in (-1, 1)$. The parameter λ_0 is determined uniquely as soon as the scattering indicatrix $g(\mu)$ has been prescribed. We recall that $\lambda = \lambda_0$ is the only nonnegative root of the transcendental equation $M(\lambda) = 1$ (Theorems 6 and 9).

Each indicatrix $g(\mu)$ is associated with its own function $M(\lambda)$ and with its own Fourier coefficient $g_0 = 2\pi \int_{-1}^{1} g(\mu)\, d\mu$. However, the quantity $M(\lambda)/g_0$ is common to the two indicatrixes, whose ratio is constant. This implies that in the class of indicatrixes differing from each other only by constant factors, the root $\lambda_0 \in [0, 1)$ of the equation $M(\lambda) = 1$ decreases monotonically with increasing g_0, assuming the value 0 for $g_0 = 1$ and approaching arbitrarily close to 1 as $g_0 \to 0 + 0$. Thus, in solving the Milne problem in the case of weak ($g_0 \approx 1$) and strong ($g_0 \ll 1$) absorption of radiation by the medium, it is important to know the behavior of $M(\lambda)$ and ϕ_λ for small values of the parameter λ, and, respectively, for values of this parameter close to unity. The present section contains derivations of the asymptotic expansions of $M(\lambda)$ and ϕ_λ in each of these limiting cases [23,26].

First let us consider the behavior of $M(\lambda)$ and ϕ_λ for small values of λ. By Theorems 7 and 15, $M(\lambda)$ and ϕ_λ are regular at the point $\lambda = 0$:

$$M(\lambda) = \sum_{\nu=0}^{\infty} M_\nu \lambda^{2\nu}, \tag{114}$$

$$\left.\begin{aligned} \varphi_\lambda &= a_\lambda + b_\lambda, \\ a_\lambda &= \sum_{\nu=0}^{\infty} \lambda^{2\nu} a^{(\nu)}, \\ b_\lambda &= \sum_{\nu=0}^{\infty} \lambda^{2\nu+1} b^{(\nu)}, \end{aligned}\right\} \tag{115}$$

(Convergence is in the metric of $C(-1, 1)$.) Here we have taken account of the evenness of $M(\lambda)$ (Theorem 9); $a^{(\nu)}$, $b^{(\nu)} \in C(-1, 1)$, $\nu = 0, 1, 2, \cdots$. The radius of the common circle of convergence of expansions (114) and (115) is positive. We denote it by R.

Let $L_2 \equiv L_2(-1, 1)$ with the scalar product $\langle f_1, f_2\rangle = \int_{-1}^{1} f_1(\mu) \overline{f_2(\mu)}\, d\mu$ and the norm $[f_1] = \langle f_1, f_1\rangle^{1/2}$, $f_{1,2} \in L_2$.

The functions ϕ_λ and $M(\lambda)$ satisfy the equation

$$M(\lambda)(1 + \lambda\mu)\varphi_\lambda(\mu) = \hat{h}_0 \varphi_\lambda(\mu), \quad \mu \in [-1, 1], \quad \lambda \in (-1, 1); \tag{116}$$

and the normalization condition

$$\langle \varphi_\lambda, 1\rangle = 2, \quad \lambda \in (-1, 1)$$

(compare Theorem 15[iii]).

This implies that $\langle a^{(0)}, 1\rangle = 2$ and $\langle a^{(\nu)}, 1\rangle = 0$ for $\nu \geq 1$, and $\langle b^{(\nu)}, 1\rangle = 0$ for $\nu \geq 0$. But $a^{(0)}(\mu) = \phi_0(\mu) = \text{const}$, so that

$$a^{(0)}(\mu) \equiv 1, \quad \mu \in [-1, 1]. \tag{117}$$

We denote by $\widetilde{L}$ the subspace of L_2 consisting of functions orthogonal to a constant. Then $a^{(\nu)} \in \widetilde{L}$, $\nu = 1, 2, \cdots$; $b^{(\nu)} \in \widetilde{L}$, $\nu = 0, 1, 2, \cdots$. Making use of Lemma 11[ii], we can readily verify that for every $f \in L_2$

$$\hat{h}_0 f(\mu) = \sum_{n=0}^{\infty} \frac{2n+1}{2} f_n g_n P_n(\mu), \tag{118}$$

where the series on the right converges uniformly with respect to all the $\mu \in [-1, 1]$; $f_n = \langle f, P_n\rangle$, g_n are the Fourier coefficients of the indicatrix $g(\mu)$: $g_n = 2\pi\langle g, P_n\rangle$ (compare Theorem 2). With the aid of formula (118) we can readily show that the operator $g_0\hat{I} - \hat{h}_0$ effects a one-to-one mapping of $\widetilde{L}$ onto itself. Since $\widetilde{L}$ is complete, this implies that $g_0\hat{I} - \hat{h}_0$, being an operator acting in $\widetilde{L}$, has a bounded inverse $\hat{\mathcal{P}}$: $\widetilde{L} \to \widetilde{L}$. If $f \in \widetilde{L}$, then

$$f = \sum_{n=1}^{\infty} \frac{2n+1}{2} f_n P_n, \quad f_n = \langle f, P_n\rangle, \quad n = 0, 1, 2, \ldots,$$

and

$$\hat{\mathcal{P}}f = \sum_{n=1}^{\infty} \frac{2n+1}{2} \cdot \frac{f_n}{g_0 - g_n} P_n. \tag{119}$$

Equation (119) implies that

$$\mathcal{P} \equiv \|\hat{\mathcal{P}}\| = \inf_{1 \leqslant n < \infty} [(g_0 - g_n)]^{-1} \geqslant g_0^{-1}.$$

Let

$$p = \mathcal{P} g_0.$$

In any case, $p \geq 1$.

We note now that

$$\varphi_{-\lambda}(\mu) = \varphi_\lambda(-\mu) \tag{120}$$

for $\lambda \in (-1, 1)$, $\mu \in [-1, 1]$ (compare Theorem 15[ii] and Theorem 11). This and formulas (115) imply that all of the $a^{(\nu)}(\mu)$ are even, and all of the $b^{(\nu)}(\mu)$ odd functions of μ. Moreover, for $|\lambda| < R$, $\mu \in [-1, 1]$ we have

$$a_\lambda(-\mu) = a_{-\lambda}(\mu) = a_\lambda(\mu),$$
$$b_\lambda(-\mu) = b_{-\lambda}(\mu) = -b_\lambda(\mu).$$

The coefficients of expansions (114) and (115) are related by recurrence relations which are readily derivable from equation (116):

$$M_0 = g_0;\ a^{(0)}(\mu) = 1,\ b^{(0)}(\mu) = -\frac{g_0}{g_0 - g_1}\mu,\ \mu \in [-1, 1]. \tag{121}$$

For $\nu \geq 1$ we have

$$\left[\sum_{k=0}^{\nu-1} M_{\nu-k} a^{(k)}(\mu) + \sum_{k=0}^{\nu-1} M_{\nu-1-k} \mu b^{(k)}(\mu)\right] \in \tilde{L},$$
$$\left[\sum_{k=0}^{\nu-1} M_{\nu-k} b^{(k)}(\mu) + \sum_{k=0}^{\nu} M_{\nu-k} \mu a^{(k)}(\mu)\right] \in \tilde{L}.$$

$$\left.\begin{aligned} M_\nu &= -\frac{1}{2} \sum_{k=0}^{\nu-1} M_{\nu-1-k} \langle b^{(k)}, \mu \rangle, \\ a^{(\nu)}(\mu) &= -\hat{\mathcal{P}}\left[\sum_{k=0}^{\nu-1} M_{\nu-k} a^{(k)}(\mu) + \sum_{k=0}^{\nu-1} M_{\nu-1-k} \mu b^{(k)}(\mu)\right], \\ b^{(\nu)}(\mu) &= -\hat{\mathcal{P}}\left[\sum_{k=0}^{\nu-1} M_{\nu-k} b^{(k)}(\mu) + \sum_{k=0}^{\nu} M_{\nu-k} \mu a^{(k)}(\mu)\right]. \end{aligned}\right\} \tag{122}$$

In fact, $M_0 = M(0) = g_0$ (Theorem 9) and $a^{(0)}(\mu) \equiv 1$ (compare (117)). Substituting expansions (114) and (115) into equation (116), we obtain

$$(g_0\hat{I} - \hat{h}_0)\, b^{(0)}(\mu) = -\, g_0\mu,$$

and, for $\nu \geq 1$,

$$(g_0\hat{I} - \hat{h}_0)\, a^{(\nu)}(\mu) = -\Big[\sum_{k=0}^{\nu-1} M_{\nu-k} a^{(k)}(\mu) + \sum_{k=0}^{\nu-1} M_{\nu-1-k}\mu b^{(k)}(\mu)\Big],$$

$$(g_0\hat{I} - \hat{h}_0)\, b^{(\nu)}(\mu) = -\Big[\sum_{k=0}^{\nu-1} M_{\nu-k} b^{(k)}(\mu) + \sum_{k=0}^{\nu} M_{\nu-k}\mu a^{(k)}(\mu)\Big].$$

Since the operator $g_0\hat{I} - \hat{h}_0$ maps L_2 onto $\widetilde{L}$, the right sides of the latter relations are contained in $\widetilde{L}$. We recall also that $a^{(\nu)} \in \widetilde{L}\,(\nu \geq 1)$ and $b^{(\nu)} \in \widetilde{L}\ (\nu \geq 0)$, from which we obtain the last two formulas of (122) and the equation

$$b^{(0)}(\mu) = -\, g_0\hat{\mathcal{P}}\mu = -\,\frac{g_0}{g_0 - g_1}\,\mu.$$

It remains for us to note that the orthogonality condition

$$\Big\langle \sum_{k=0}^{\nu-1} M_{\nu-k} a^{(k)}(\mu) + \sum_{k=0}^{\nu-1} M_{\nu-1-k}\mu b^{(k)}(\mu),\ 1\Big\rangle = 0$$

is equivalent to the first formula of (122).

Making use of formulas (121) and (122), we can obtain a lower estimate of the radius of convergence of expressions (114) and (115) and investigate the structure of the coefficients $a^{(\nu)}$ and $b^{(\nu)}$. This was done in [23]. It turns out that R admits of a simple minorant which depends on the parameter p introduced above:

$$R \geqslant R_0(p),$$

$$R_0(p) = \frac{1}{p}\left\{\frac{n^2 - 3 - (n+1)\sqrt{(n-3)(n+1)}}{6\left(1 - \dfrac{1}{p\sqrt{3}}\right)}\right\}^{1/2},$$

$$n = 3\left(1 + \frac{\sqrt{3}}{2}\,\frac{p+1}{p - 1/\sqrt{3}}\right).$$

As regards the quantities M_ν, $a^{(\nu)}$, $b^{(\nu)}$, these can be computed from recurrence formulas (121) and (122). It turns out here that $a^{(\nu)}$ and $b^{(\nu)}$ are polynomials of degree 2ν and $2\nu + 1$ (respectively) in the variable μ, and that their coefficients are related by simple expressions.

Let $N(\lambda) = g_0^{-1}M(\lambda)$, so that for $|\lambda| < R$ we have

$$N(\lambda) = \sum_{\nu=0}^{\infty} N_\nu \lambda^{2\nu}, \quad N_\nu = g_0^{-1}M_\nu, \quad \nu = 0, 1, 2, \ldots.$$

Let us set

$$q_k = [1 - g_k g_0^{-1}]^{-1}, \quad k = 1, 2, \ldots.$$

It is clear that $q_k > 0$ for all $k \geq 1$.

Proceeding from (121), (122) and from the explicit form of the operator $\hat{\mathcal{P}}$, we can obtain the following result [23] by induction over ν:

For $\mu \in [-1, 1]$,

$$\left.\begin{aligned} a_0(\mu) &= A_0^{(0)}, \\ a^{(\nu)}(\mu) &= \sum_{\tau=1}^{\nu} A_\tau^{(\nu)} P_{2\tau}(\mu), \quad \nu \geqslant 1; \\ b^{(\nu)}(\mu) &= \sum_{\tau=0}^{\nu} B_\tau^{(\nu)} P_{2\tau+1}(\mu), \quad \nu \geqslant 0. \end{aligned}\right\} \tag{123}$$

The numbers N_ν, $A_\tau^{(\nu)}$, $B_\tau^{(\nu)}$ are determined uniquely by specification of all the q_k, $k \geq 1$. They obey the following relations, from which they can be readily computed in succession:

$$N_0 = 1, \quad A_0^{(0)} = 1, \quad B_0^{(0)} = -q_1,$$

$$N_1 = \frac{1}{3} q_1, \quad A_1^{(1)} = \frac{2}{3} q_1 q_2, \quad B_0^{(1)} = -\frac{1}{3} q_1^2 \left(1 - q_1 + \frac{4}{5} q_2\right),$$

$$B_1^{(1)} = -\frac{2}{5} q_1 q_2 q_3,$$

$$A_\nu^{(\nu)} = \frac{(2\nu)!}{(4\nu - 1)!!} \prod_{s=1}^{2\nu} q_s, \quad \nu \geqslant 1,$$

$$B_\nu^{(\nu)} = -\frac{(2\nu + 1)!}{(4\nu + 1)!!} \prod_{s=1}^{2\nu+1} q_s, \quad \nu \geqslant 0.$$

For $\nu \geq 2$,

$$N_\nu = -\frac{1}{3} \sum_{k=0}^{\nu-1} N_{\nu-1-k} B_0^{(k)},$$

$$A_\tau^{(\nu)} = -q_{2\tau} \left\{ \sum_{k=\tau}^{\nu-1} N_{\nu-k} A_\tau^{(k)} + \frac{2\tau}{4\tau - 1} \sum_{k=\tau-1}^{\nu-1} N_{\nu-1-k} B_{\tau-1}^{(k)} \right.$$
$$\left. + \frac{2\tau + 1}{4\tau + 3} \sum_{k=\tau}^{\nu-1} N_{\nu-1-k} B_\tau^{(k)} \right\}; \quad 1 \leqslant \tau \leqslant \nu - 1;$$

$$B_0^{(\nu)} = -q_1 \left[N_\nu (1 - q_1) + \sum_{k=1}^{\nu-1} N_{\nu-k} B_0^{(k)} + \frac{2}{5} \sum_{k=1}^{\nu} N_{\nu-k} A_1^{(k)} \right],$$

$$B_\tau^{(\nu)} = -q_{2\tau+1}\left\{\sum_{k=\tau}^{\nu-1} N_{\nu-k}B_\tau^{(k)} + \frac{2\tau+1}{4\tau+1}\sum_{k=\tau}^{\nu} N_{\nu-k}A_\tau^{(k)} + \frac{2\tau+2}{4\tau+5}\sum_{k=\tau+1}^{\nu} N_{\nu-k}A_{\tau+1}^{(k)}\right\}, \quad 1 \leqslant \tau \leqslant \nu-1.$$

From expansions (123) we see that the higher angular harmonics of the function $\phi_\lambda(\mu)$ vanish as $\lambda \to 0$ the faster the higher their number. In fact, it is easy to verify that the scalar product $\langle\phi_\lambda, P_k\rangle$ has a zero of order k at the point $\lambda = 0$. We set

$$\psi_k(\lambda) = \frac{2k+1}{2}\lambda^{-k}\langle\varphi_\lambda, P_k\rangle, \tag{124}$$

$k = 0, 1, 2, \cdots;\ \lambda \in G_0$ (compare Theorem 15).

The functions $\psi_k(\lambda)$ are holomorphic in the domain G_0, and

$$\psi_k(0) = (-1)^k \frac{k!}{(2k-1)!!}\prod_{s=1}^{k} q_s \neq 0, \quad k = 1, 2, \ldots. \tag{125}$$

Making use of (120), we can readily verify that all of the $\psi_k(\lambda)$ are even functions of the variable λ.

Now let us investigate the behavior of $M(\lambda)$ and ϕ_λ for λ close to unity. In this case (and only in the remainder of this section) we shall assume that the scattering indicatrix is a polynomial of finite degree n_0, i.e. that

$$g(\mu) = \sum_{n=0}^{n_0}\frac{2n+1}{4\pi}g_nP_n(\mu) \geqslant 0, \quad \mu \in [-1, 1],$$

and that

$$g(1) > 0.$$

In these assumptions the kernel of the operator $\hat{h}_0$ (compare §5) turns out to be degenerate:

$$h_0(\mu, \mu') = \sum_{n=0}^{n_0}\frac{2n+1}{2}g_nP_n(\mu)P_n(\mu').$$

Let

$$\alpha_i(\lambda) = \langle\varphi_\lambda, P_i\rangle,$$

$$c_{ij}^0(\lambda) = \int_1^1 \frac{P_i(\mu)P_j(\mu)}{1+\lambda\mu}d\mu,$$

$$i, j = 0, 1, 2, \ldots, n_0$$

for $\lambda \in (-1, 1)$. Then (compare (116))

$$\varphi_\lambda(\mu) = \frac{1}{M(\lambda)(1+\lambda\mu)} \sum_{k=0}^{n_0} \frac{2k+1}{2} g_k \alpha_k(\lambda) P_k(\mu),$$

$$M(\lambda)\alpha_i(\lambda) = \sum_{j=0}^{n_0} \frac{2j+1}{2} g_j c_{ij}^0(\lambda)\alpha_j(\lambda), \quad i = 0, 1, 2, \ldots, n_0.$$

We set

$$\widetilde{M}(\lambda) = \left[\frac{1}{\lambda}\ln\frac{1+\lambda}{1-\lambda}\right]^{-1} M(\lambda),$$

$$c_{ij}(\lambda) = \left[\frac{1}{\lambda}\ln\frac{1+\lambda}{1-\lambda}\right]^{-1} c_{ij}^0(\lambda),$$

$$\alpha(\lambda) = \{\alpha_0(\lambda), \alpha_1(\lambda), \ldots, \alpha_{n_0}(\lambda)\}.$$

The vector $\alpha(\lambda)$ is a solution of the matrix equation

$$\widetilde{M}(\lambda)\alpha(\lambda) = A(\lambda)\alpha(\lambda), \tag{126}$$

where $A(\lambda) = C(\lambda)H$ and $C(\lambda)$ is a real symmetric matrix with the elements $c_{ij}(\lambda)$; H is a diagonal matrix with the elements $h_{ij} = h_j\delta_{ij}$, $h_j = (2j+1)g_j/2$. It follows from Theorem 6 that $\widetilde{M}(\lambda)$ is the eigenvalue of largest absolute value of equation (126).

Let us associate the matrix $C(\lambda)$ with the matrix $C(\epsilon, \delta)$ whose elements $c_{ij}(\epsilon, \delta)$ are entire functions of the pair of variables ϵ, δ:

$$c_{ij}(\varepsilon, \delta) = \sum_{m=0}^{i+j} b_{ij}^{(m)}\delta^m + \varepsilon \sum_{n=0}^{i+j-1} d_{ij}^{(n)}\delta^n.$$

This form $c_{ij}(\epsilon, \delta)$ and the real numbers $b_{ij}^{(m)}$ and $d_{ij}^{(n)}$ are chosen in such a way [26] that

$$C\left(\frac{1}{\ln\frac{1+\lambda}{1-\lambda}}, \frac{1}{\lambda} - 1\right) = C(\lambda), \quad \lambda \in (0, 1).$$

If the number $\widetilde{M}(\epsilon, \delta)$ and the vector $x(\epsilon, \delta)$ now satisfy the matrix equation

$$A(\varepsilon, \delta)x(\varepsilon, \delta) = \widetilde{M}(\varepsilon, \delta)x(\varepsilon, \delta), \tag{127}$$

where $A(\epsilon, \delta) = C(\epsilon, \delta)H$, then the pair $\overline{\widetilde{M}}(\lambda)$, $\overline{\alpha}(\lambda)$, given by

$$\overline{\widetilde{M}}(\lambda) = \widetilde{M}\left(\frac{1}{\ln\frac{1+\lambda}{1-\lambda}}, \frac{1}{\lambda} - 1\right),$$

$$\overline{\alpha}(\lambda) = x\left(\frac{1}{\ln\frac{1+\lambda}{1-\lambda}}, \frac{1}{\lambda} - 1\right)$$

is the solution of equation (126).

$A(\epsilon, \delta)$ is an analytic perturbation of the matrix $A_{00} = A(0, 0)$:

$$A(\varepsilon, \delta) = \sum_{\nu=0}^{2n_0} \delta^\nu A_{0\nu} + \varepsilon \sum_{\nu=0}^{2n_0-1} \delta^\nu A_{1\nu}.$$

Here

$$A_{0\nu} = b^{(\nu)} H, \quad A_{1\nu} = d^{(\nu)} H,$$

and $b^{(\nu)}$ and $d^{(\nu)}$ are matrices with the elements $b_{ij}^{(\nu)}$ and $d_{ij}^{(\nu)}$, respectively.

On the basis of the unperturbed equation

$$A_{00} x = kx,$$

we can use the standard methods of perturbation theory to construct a solution of (127) in the neighborhood of the point $(\epsilon, \delta) = (0, 0)$ which is reduced by the substitution

$$\varepsilon = \left(\ln \frac{1+\lambda}{1-\lambda}\right)^{-1}, \ \delta = \frac{1}{\lambda} - 1$$

to the required solution $\widetilde{M}(\lambda)$, $\alpha(\lambda)$ of equation (126). The appropriate operations are carried out in [26]. We shall merely cite the final results here.

There exists a positive number ρ_0 (for which we can obtain a lower estimate) such that for $\lambda \in (1 - \rho_0, 1)$ we have

$$\left.\begin{aligned} \widetilde{M}(\lambda) &= \sum_{m, n=0}^{\infty} \widetilde{M}_{mn} \left[\ln \frac{1+\lambda}{1-\lambda}\right]^{-m} \left(\frac{1}{\lambda} - 1\right)^n, \\ \alpha(\lambda) &= \sum_{m, n=0}^{\infty} \left[\ln \frac{1+\lambda}{1-\lambda}\right]^{-m} \left(\frac{1}{\lambda} - 1\right)^n x_{mn}. \end{aligned}\right\} \tag{128}$$

The series on the right converge absolutely and uniformly with respect to $\lambda \in (1 - \rho_0, 1)$. The numerical and vector coefficients $\widetilde{M}_{mn} x_{mn}$ of these series obey the following recurrence relations, from which they can be computed in succession:

$$\widetilde{M}_{00} = 2\pi g(1), \quad x_{00} = \{1, -1, 1, -1, \ldots, (-1)^{n_0}\}.$$

$$\widetilde{M}_{mn} = \frac{1}{2\pi g(1)} \left[\sum_{k=1}^{n} (A_{0k} x_{m, n-k}, \overset{*}{x}_{00}) + \sum_{k=0}^{n} (A_{1k} x_{m-1, n-k}, \overset{*}{x}_{00})\right],$$

$$\begin{aligned} x_{mn} = \frac{1}{2\pi g(1)} \Bigg[& \sum_{k=1}^{n} A_{0k} x_{m, n-k} + \sum_{k=0}^{n} A_{1k} x_{m-1, n-k} \\ & - \sum_{m'=0}^{m-1} \widetilde{M}_{m-m', 0} x_{m', n} - \sum_{m'=0}^{m} \sum_{k=0}^{n-1} \widetilde{M}_{m-m', n-k} x_{m', k} \Bigg], \end{aligned}$$

$$m = 0, 1, 2, \ldots, n = 0, 1, 2, \ldots, (m, n) \neq (0, 0).$$

Here $x^*_{00} = \{h_0, -h_1, h_2, \cdots, (-1)^{n_0} h_{n_0}\}$; (x, y) is the scalar product of the vectors $x = \{x_0, \cdots, x_{n_0}\}$ and $y = \{y_0, \cdots, y_{n_0}\}$: $(x, y) = \sum_{k=0}^{n_0} x_k y_k$ · $x_{-1,\nu} = 0$ $(\nu \geq 0)$,

$$A_{0\nu} = 0 \text{ for } \nu > 2n_0, \quad A_{1\nu} = 0 \text{ for } \nu > 2n_0 - 1.$$

In these and subsequent formulas we assume that $\sum_{q=a}^{b} \cdots = 0$, if $b < a$.

Isolating the principal terms in (128), we arrive at the following expressions:

$$\widetilde{M}(\lambda) = 2\pi g(1) - \frac{1}{\pi g(1)} \Big[\sum_{i,\,j=0}^{n_0} h_i h_j \sum_{\nu=1}^{\max\{i,\,j\}} \frac{1}{\nu} \Big] \Big(\ln \frac{1+\lambda}{1-\lambda} \Big)^{-1} + O\Big(\Big[\ln \frac{1+\lambda}{1-\lambda} \Big]^{-2} \Big),$$

$$\alpha(\lambda) = x_{00} + \frac{1}{2\pi g(1)} \Big\{ A_{10} + \frac{1}{\pi g(1)} \Big[\sum_{i,\,j=0}^{n_0} h_i h_j \sum_{\nu=1}^{\max\{i,\,j\}} \frac{1}{\nu} \Big] E \Big\} \Big(\ln \frac{1+\lambda}{1-\lambda} \Big)^{-1} x_{00}$$
$$+ O\Big(\Big[\ln \frac{1+\lambda}{1-\lambda} \Big]^{-2} \Big).$$

(Here E is an unit matrix of order $n_0 + 1$.)

As already noted, with large absorptances (i.e. with small g_0) the root λ_0 of the equation $M(\lambda) = 1$ is close to unity, and we can make use of the above expansions. The directional distribution of the radiation in the deep portions of the halfspace $\tau > 0$ is defined by the function

$$\varphi_{\lambda 0}(\omega \mathbf{n}) \approx \frac{1}{1 + \lambda_0 \omega \mathbf{n}} \sum_{k=0}^{n_0} \frac{2k+1}{2} g_k (-1)^k P_k(\omega \mathbf{n}) = 2\pi \frac{g(-\omega \mathbf{n})}{1 + \lambda_0 \omega \mathbf{n}}.$$

According to this formula, the deep scattering state in high-absorptance media can be used for approximate reconstruction of the scattering indicatrix. In the next section we shall express this result in strict form. Specifically, we shall show that the asymptotic characteristics of the radiation density in the nonconservative case determine the scattering indicatrix uniquely.

§13. The inverse problem of the asymptotic theory of radiation transfer [24,25]

In §3 we agreed to apply the term "indicatrix" to functions defined on the segment $[-1, 1]$ which satisfy Condition I (q.v.). We denote by Γ the set of all indicatrixes for each of which $g_0 < 1$. Thus Γ contains all of the indicatrixes corresponding to nonconservative scattering laws.

Let $g \in \Gamma$ and $\lambda \in (-1, 1)$. On the basis of Theorems 6, 15 and 13 we can readily show that the operator $(1 + \lambda\mu)^{-1} \hat{h}_0$ has in $L_2 = L_2(-1, 1)$ exactly one nonnegative eigenfunction u which satisfies the normalization condition $\langle u, 1 \rangle = 2$.

This function corresponds to the simple positive eigenvalue $M_g(\lambda) \equiv M(\lambda)$, where $M(\lambda)$ is the quantity defined in Theorem 6; it coincides with the function ϕ_λ (see Theorem 15). The properties of $M_g(\lambda)$ and ϕ_λ were investigated in §§4 and 5.

We denote the correspondence $g \to \phi_\lambda$ by the symbol $\check{\omega}$: $\phi_\lambda = \check{\omega} g$. Thus, for every $g \in \Gamma$, $\check{\omega} g$ is some uniquely defined abstract function of the variable $\lambda \in (-1, 1)$ with values in $C(-1, 1)$. We denote the collection of all functions $\check{\omega} g$, $g \in \Gamma$, by Ψ,

$$\Psi = \{\check{\omega} g \mid g \in \Gamma\}.$$

By definition, $\check{\omega}$ is a single-valued mapping of Γ onto Ψ.

Let $g \in \Gamma$ and $\phi_\lambda = \check{\omega} g$. There exists exactly one $\lambda_0 = \lambda_0(g) \in (0, 1)$ such that $M_g(\lambda_0) = 1$ (Theorem 9). We set $\phi^{[g]} = \phi_{\lambda_0(g)}$. Specifying $g \in \Gamma$ determines $\phi^{[g]}$ uniquely, $\phi^{[g]} \in C(-1, 1)$.

The pair $\lambda_0(g)$, $\phi^{[g]}$ characterizes the asymptotic behavior of the radiation distribution at large depths in the homogeneous Milne problem with a given indicatrix g:

$$I(\tau, \omega) \sim e^{\tau\lambda_0(g)} \varphi^{[g]}(\omega \mathbf{n}).$$

The converse also turns out to be true: the prescribed deep radiation scattering state (i.e. $\lambda_0(g)$ and $\phi^{[g]}$) determines the indicatrix uniquely. Moreover, as is evident from Theorem 31 below, the indicatrix g and the parameter $\lambda_0(g)$ are determined uniquely once we specify the function $\phi^{[g]}$ alone.

Let us denote the above correspondence $g \to \phi^{[g]}$ by $\check{\omega}_0$,

$$\check{\omega}_0 g = \phi^{[g]};$$

by Ψ_0 we denote the class of all functions $\phi^{[g]}$ which arises when g runs through the set Γ,

$$\Psi_0 = \{\check{\omega}_0 g \mid g \in \Gamma\}.$$

By construction, $\check{\omega}_0$ is a single-valued mapping of Γ onto Ψ_0.

Theorem 31. *$\check{\omega}_0$ effects a one-to-one mapping of Γ onto Ψ_0.*

This theorem can be proved by partial description of the classes Ψ and Ψ_0. Before turning to this matter, we note that Theorem 31 ceases being valid if the class Γ is made to include indicatrixes corresponding to conservative scattering laws ($g_0 = 1$). In fact, in the conservative case the directional distribution of radiation at large optical depths is always isotropic, whatever the scattering indicatrix (compare §11).

We choose an arbitrary function $\phi_\lambda \in \Psi$. There exists a $g \in \Gamma$ such that $\phi_\lambda = \check{\omega} g$. This means that

$$M_g(\lambda)(1+\lambda\mu)\varphi_\lambda(\mu) = \int_{-1}^{1} h_0(\mu,\mu')\varphi_\lambda(\mu')\,d\mu', \tag{129}$$

for all $\mu \in [-1, 1]$ and $\lambda \in (-1, 1)$.

The kernel $h_0(\mu, \mu')$ is related to the indicatrix g by the formulas of Lemma 11. Being a continuous function of μ, the function $\phi_\lambda(\mu)$ can be expanded in a series in Legendre polynomials which converges at least in the mean for each fixed value $\lambda \in (-1, 1)$,

$$\varphi_\lambda(\mu) = \sum_{k=0}^{\infty} \lambda^k \psi_k(\lambda) P_k(\mu). \tag{130}$$

The functions $\psi_k(\lambda)$ are defined by relation (124) and are holomorphic in the neighborhood G_0 of the interval $(-1, 1)$ (Theorem 15). For what follows it is especially important to note that

$$\psi_k(0) \neq 0, \quad \operatorname{sgn} \psi_k(0) = (-1)^k, \quad k = 0, 1, 2, \ldots$$

(compare (125)). We also note that for $\lambda \in G_0$,

$$\psi_0(\lambda) = \frac{1}{2}\langle \varphi_\lambda, 1\rangle = 1.$$

Substituting (130) into (129), we arrive at an infinite system of equations relating the $\psi_k(\lambda)$:

$$\frac{k}{2k-1}\psi_{k-1}(\lambda) + \left[1 - \frac{g_k}{M_g(\lambda)}\right]\psi_k(\lambda) + \lambda^2 \frac{k+1}{2k+3}\psi_{k+1}(\lambda) = 0, \tag{131}$$

$$\psi_{-1}(\lambda) \equiv 0, \quad k = 0, 1, 2, \ldots, \lambda \in (-1, 1).$$

We shall investigate system (131) by the methods of continued fractions. We set

$$b_k(\lambda) = -\frac{2k+1}{k+1}\left(1 - \frac{g_{k+1}}{M_g(\lambda)}\right), \quad k \geqslant 0;$$

$$a_k(\lambda) = -\lambda^2 \frac{k+1}{k} \cdot \frac{2k-1}{2k+3}, \quad k \geqslant 1.$$

System (131) (without the first equation) can now be written as

$$\psi_k(\lambda) = b_k(\lambda)\psi_{k+1}(\lambda) + a_{k+1}(\lambda)\psi_{k+2}(\lambda). \quad k \geqslant 0. \tag{132}$$

By Theorem 7, $b_k(\lambda)$ and $a_k(\lambda)$ are analytic in the domain G_0. Hence relations (132) are fulfilled at all points λ of this domain.

To each number $\sigma \in (0, 1)$ we assign a positive number η_σ in such a way that the closed set

$$S_\sigma \equiv \{\lambda \,|\, |\lambda| \leqslant \sigma, \ |\operatorname{Im} \lambda| \leqslant \eta_\sigma\}$$

is contained entirely in G_0. The resulting family of sets S_σ, $\sigma \in (0, 1)$ has the following self-evident property:

$$(-1, 1) \subset \bigcup_{\sigma \in (0,1)} S_\sigma \subset G_0.$$

It is easy to see that

$$\lim_{k=\infty} b_k(\lambda) = -2, \quad \lim_{k=\infty} a_k(\lambda) = -\lambda^2 \tag{133}$$

uniformly with respect to all $\lambda \in S_\sigma$. We now associate system (132) with the continued fractions ($k \geq 0$)

$$\xi_k(\lambda) = b_k(\lambda) + \cfrac{a_{k+1}(\lambda)}{b_{k+1}(\lambda) + \cfrac{a_{k+2}(\lambda)}{b_{k+2}(\lambda) + \dots}}.$$

The quantity $\xi_k(\lambda)$ is the limit (provided such a limit exists) of the sequence of convergents

$$\xi_k(\lambda) = \lim_{\nu=\infty} \frac{A_\nu^{(k)}(\lambda)}{B_\nu^{(k)}(\lambda)}. \tag{134}$$

The numerator and denominator of the convergent can be found from the familiar recurrence relations

$$A_0^{(k)}(\lambda) = b_k(\lambda), \quad A_1^{(k)}(\lambda) = b_k(\lambda)\, b_{k+1}(\lambda) + a_{k+1}(\lambda);$$
$$B_0^{(k)}(\lambda) = 1, \qquad B_1^{(k)}(\lambda) = b_{k+1}(\lambda);$$

and for $\nu \geq 2$ we have

$$A_\nu^{(k)}(\lambda) = b_{k+\nu}(\lambda)\, A_{\nu-1}^{(k)}(\lambda) + a_{k+\nu}(\lambda)\, A_{\nu-2}^{(k)}(\lambda);$$
$$B_\nu^{(k)}(\lambda) = b_{k+\nu}(\lambda)\, B_{\nu-1}^{(k)}(\lambda) + a_{k+\nu}(\lambda)\, B_{\nu-2}^{(k)}(\lambda).$$

Convergence of the sequence of convergents depends on the properties of the roots $\rho_{1,2}(\lambda)$ of the quadratic equation whose coefficients are determined by limits (133):

$$\rho^2 + 2\rho + \lambda^2 = 0.$$

We have

$$\rho_1(\lambda) = -(1 + \sqrt{1-\lambda^2}), \quad \rho_2(\lambda) = -(1 - \sqrt{1-\lambda^2}).$$

It is easy to see that, for $\lambda \in S_\sigma$ and $\sigma \in (0, 1)$,

a) $1 < |\rho_1(\lambda)| < 3$,

b) $|\rho_2(\lambda)/\rho_1(\lambda)| \leq \sigma^2 < 1$.

This is already sufficient to enable us to use the methods of continued fraction theory (see for example Theorems 40, 41, 42 of [40]) to investigate the convergence of the fractions $\xi_k(\lambda)$ for large k.

Lemma 33. *For every* $\sigma \in (0, 1)$ *there exists a natural number* $k(\sigma)$ *which has the following properties:*

[i] *If* $k \geq k(\sigma)$, *then there exists a* $\nu_0 = \nu_0(k, \sigma)$ *such that the denominator* $B_\nu^{(k)}(\lambda)$ *is not equal to zero for* $\nu \geq \nu_0$, $\lambda \in S_\sigma$.

[ii] *For* $k \geq k(\sigma)$ *limit relation* (134) *is fulfilled uniformly with respect to all* $\lambda \in S_\sigma$.

[iii] *The functions* $\xi_k(\lambda)$, $k \geq k(\sigma)$, *are analytic in* S_σ.

[iv]
$$\lim_{k=\infty} \xi_k(\lambda) = \rho_1(\lambda)$$

uniformly with respect to all $\lambda \in S_\sigma$.

The coefficients $\psi_k(\lambda)$ and the continued fractions $\xi_k(\lambda)$ are related in a simple way. This is clear in the case $\lambda = 0$, where $a_{k+1}(0) = 0$, $b_k(0) = \xi_k(0)$, and equations (132) become

$$\psi_k(0) = \xi_k(0)\, \psi_{k+1}(0).$$

This remains valid for $\lambda \neq 0$, at least for sufficiently large k.

Let $\lambda \in (-1, 1)$, $\lambda \neq 0$. We have

$$|\psi_k(\lambda)| = \frac{2k+1}{2|\lambda|^k} |\langle \varphi_\lambda, P_k \rangle| \leqslant \frac{2k+1}{2|\lambda|^k} \langle \varphi_\lambda, 1 \rangle = \frac{2k+1}{|\lambda|^k}.$$

Hence

$$\limsup_{k=\infty} |\psi_k(\lambda)|^{1/k} \leqslant \frac{1}{|\lambda|}.$$

At the same time, for $\lambda \in (-1, 1)$, $\lambda \neq 0$, we have

$$\frac{1}{|\rho_2(\lambda)|} = \frac{1}{1 - \sqrt{1-\lambda^2}} = \frac{1 + \sqrt{1-\lambda^2}}{\lambda^2} > \frac{1}{|\lambda|}.$$

Thus

c) $\lim_{k=\infty} \sup |\psi_k(\lambda)|^{1/k} < 1/|\rho_2(\lambda)|$, $\lambda \neq 0$, $\lambda \in (-1, 1)$.

Let us recall one further fact, i.e. that

d) $a_k(\lambda) \neq 0$ for $\lambda \in (-1, 1)$, $\lambda \neq 0$, $k = 1, 2, \cdots$.

Lemma 34. *Let* $\sigma \in (0, 1)$. *For* $k \geq k(\sigma)$ *and* $\lambda \in (-\sigma, \sigma)$ *we have*

$$\psi_k(\lambda) = \xi_k(\lambda)\, \psi_{k+1}(\lambda).$$

This lemma follows from conditions a) – d) and from Lemma 33. It can be proved by the usual devices of the theory of continued fractions (its proof is essentially similar to the proof of Theorem 46 of [40]).

Corollary. *$\psi_k(\lambda) \neq 0$ for $k > k(\sigma)$ and $\lambda \in (-\sigma, \sigma)$.*

In fact, if $\psi_{k_1}(\lambda_1) = 0$, $k_1 > k(\sigma)$, $\lambda_1 \in (-\sigma, \sigma)$, then by Lemma 34 it is also the case that $\psi_{k_1-1}(\lambda_1) = 0$. But (132) then implies that $\psi_k(\lambda_1) = 0$ for all $k \leq k_1$, including $\psi_0(\lambda_1) = 0$, which is not true.

Lemma 35. *Let $\sigma \in (0, 1)$, $k_0 \geq 0$, $\psi_{k_0}(\lambda) \neq 0$ for $\lambda \in (-\sigma, \sigma)$. Then $\psi_k(\lambda) \neq 0$ for $0 \leq k \leq k_0$ and $\lambda \in (-\sigma, \sigma)$.*

This can be proved by induction over k_0. For $k_0 = 0$ and $k_0 = 1$ the statement is self-evident, since $\psi_0(\lambda) = 1$. Let the lemma be valid for $k_0 \leq n$, $n \geq 1$, and for an arbitrary $\sigma \in (0, 1)$. Let us prove it for $k_0 = n + 1$.

We have $\psi_{n+1}(\lambda) \neq 0$ for $\lambda \in (-\sigma, \sigma)$. Let us assume that $\psi_n(\lambda)$ has zeros in the interval $(-\sigma, \sigma)$. Since $\psi_n(\lambda)$ is even and $\psi_n(0) \neq 0$, the set $\mathfrak{N}$ of zeros of $\psi_n(\lambda)$ belonging to $(0, \sigma)$ is nonempty.

Let $\sigma_1 = \inf \mathfrak{N}$. Then $\psi_n(\sigma_1) = 0$, $\sigma_1 \in (0, \sigma)$, and $\psi_n(\lambda) \neq 0$ for $\lambda \in (-\sigma_1, \sigma_1)$. By the inductive assumption this implies that $\psi_{n-1}(\lambda) \neq 0$ for $\lambda \in (-\sigma_1, \sigma_1)$.

On the other hand, turning to (131), we find that

$$\frac{n}{2n-1}\psi_{n-1}(\sigma_1) + \sigma_1^2\frac{n+1}{2n+3}\psi_{n+1}(\sigma_1) = 0.$$

$\psi_{n+1}(\sigma_1) \neq 0$, so that $\psi_{n-1}(\sigma_1) \neq 0$. Moreover, $\operatorname{sgn}\psi_{n-1}(\sigma_1) = -\operatorname{sgn}\psi_{n+1}(\sigma_1)$. But $[0, \sigma_1] \subset (-\sigma, \sigma)$, so that $\psi_{n+1}(\lambda)$ does not vanish on $[0, \sigma_1]$. Hence $\operatorname{sgn}\psi_{n+1}(\sigma_1) = \operatorname{sgn}\psi_{n+1}(0) = \operatorname{sgn}\psi_{n-1}(0)$ (compare (125)). Thus $\operatorname{sgn}\psi_{n-1}(\sigma_1) = -\operatorname{sgn}\psi_{n-1}(0)$, which means that $\psi_{n-1}(\lambda)$ vanishes at some point of the interval $(0, \sigma_1)$.

This contradiction means that $\psi_n(\lambda)$ does not have zeros in the interval $(-\sigma, \sigma)$. By the inductive assumption this implies that $\psi_k(\lambda)$ also does not vanish on $(-\sigma, \sigma)$, $0 \leq k \leq n$. The lemma is proved.

Theorem 32. *Let $\varphi_\lambda \in \Psi$ and*

$$\psi_k(\lambda) = \frac{2k+1}{2\lambda^k}\langle\varphi_\lambda, P_k\rangle, \quad k = 0, 1, 2, \ldots, \lambda \in (-1, 1).$$

The functions $\psi_k(\lambda)$ are regular, real, and different from zero at every point of the interval $(-1, 1)$, $k = 0, 1, 2, \cdots$;

$$\operatorname{sgn}\psi_k(\lambda) = (-1)^k, \quad \lambda \in (-1, 1), \; k = 0, 1, 2, \ldots.$$

For every $\sigma \in (0, 1)$,

$$\lim_{k=\infty}\frac{\psi_k(\lambda)}{\psi_{k+1}(\lambda)} = -(1 + \sqrt{1-\lambda^2})$$

uniformly with respect to $\lambda \in (-\sigma, \sigma)$.

Proof. Let $\sigma \in (0, 1)$. By the corollary of Lemma 34, $\psi_k(\lambda)$ does not vanish on $(-\sigma, \sigma)$ for $k \geq k(\sigma) + 1$. Choosing a $k_0 = k(\sigma) + 1$ and making use of Lemma 35, we find from this that $\psi_k(\lambda) \neq 0$ on $(-\sigma, \sigma)$ for $0 \leq k \leq k_0$. Thus all of the $\psi_k(\lambda)$ are different from zero on $(-\sigma, \sigma)$. But σ is an arbitrary point of (0, 1), which means that

$$\psi_k(\lambda) \neq 0, \ \lambda \in (-1, 1), \ k = 0, 1, 2, \ldots.$$

It was stated above that the functions $\psi_k(\lambda)$ are holomorphic on some neighborhood G_0 of the interval $(-1, 1)$. (This neighborhood is determined as soon as we are given a $g \in \Gamma$ satisfying the condition $\check{\omega} g = \phi_\lambda$; cf. Theorem 15.) Thus the $\psi_k(\lambda)$ are obviously regular at points of the interval $(-1, 1)$, and therefore

$$\operatorname{sgn} \psi_k(\lambda) = \operatorname{sgn} \psi_k(0) = (-1)^k, \quad \lambda \in (-1, 1), \ k \geqslant 0.$$

The limit relation follows from Lemmas 34 and 33[iv]. The theorem is proved.

The above results enable us to discern some of the properties of the elements of the class Ψ_0.

Theorem 33. *Let $\phi \in \Psi_0$ be a positive continuous function of $\mu \in [-1, 1]$, $\langle \phi, 1 \rangle = 2$. Denote by ϕ_k the coefficients of the expansion of ϕ in a series in Legendre polynomials,*

$$\varphi_k = \frac{2k+1}{2} \langle \varphi, P_k \rangle, \quad k = 0, 1, 2, \ldots.$$

Then the following assertions are valid.

[i] *The series*

$$\sum_{k=0}^{\infty} \varphi_k P_k(\mu) = \varphi(\mu)$$

converges uniformly with respect to all $\mu \in [-1, 1]$.

[ii] *The coefficients ϕ_k form an alternating-sign sequence all of whose elements differ from zero:*

$$\varphi_0 = 1; \quad \varphi_k \neq 0, \quad \operatorname{sgn} \varphi_k = (-1)^k, \quad k = 0, 1, 2, \ldots$$

[iii] *The limit*

$$\lim_{k=\infty} \frac{\varphi_{k-1}}{\varphi_k} = -\alpha$$

exists. Here $\alpha > 1$, and

$$\frac{\varphi_{k-1}}{\varphi_k} = -\alpha\left(1 - \frac{1}{2k}\right) + \delta_k, \ k = 1, 2, \ldots, \tag{135}$$

where

$$\sum_{k=1}^{\infty} k\delta_k^2 < \infty. \tag{136}$$

[iv] *Let* $g \in \Gamma$ *and* $\check{\omega}_0 g = \phi$. *Set* $\phi_\lambda = \check{\omega} g$. *Then*

$$\lambda_0(g) = 2\alpha(\alpha^2+1)^{-1}, \quad \varphi = \varphi_{\lambda_0(g)}$$

and

$$g_k = 1 + \lambda_0(g)\left[\frac{k}{2k-1}\frac{\varphi_{k-1}}{\varphi_k} + \frac{k+1}{2k+3}\frac{\varphi_{k+1}}{\varphi_k}\right], \quad k \geqslant 0, \ \varphi_{-1} = 0. \tag{137}$$

Proof. Every function $\phi \in \Psi_0$ can be expressed in the form $\phi = \check{\omega}_0 g$, $g \in \Gamma$. By the definition of $\check{\omega}_0$ we have

$$\varphi = \varphi^{[g]} = \varphi_{\lambda_0(g)}.$$

The properties of the functions ϕ_λ are described in Theorem 15, which implies that $\phi = \phi_{\lambda_0(g)} \in (-1, 1)$, $\phi(\mu) > 0$ at all points of the segment $[-1, 1]$, and that $\langle\phi, 1\rangle = 2$.

We choose a $\phi \in \Psi_0$ and set

$$\varphi_k = \frac{2k+1}{2}\langle\varphi, P_k\rangle, \quad k = 0, 1, 2, \ldots.$$

We can now express ϕ as a series

$$\varphi = \sum_{k=0}^{\infty} \varphi_k P_k,$$

which converges in the metric of L_2.

We choose a $g \in \Gamma$ such that $\check{\omega}_0 g = \phi$ and set $\phi_\lambda = \check{\omega} g$. It is clear that $\phi_\lambda \in \Psi$, $\phi(\mu) = \phi_{\lambda_0(g)}(\mu)$, and that $\lambda = \lambda_0(g)$ is defined by the condition $M_g(\lambda) = 1$, $0 < \lambda_0(g) < 1$.

We associate the quantities $\psi_k(\lambda)$ with the function ϕ_λ in accordance with formula (124). The relations

$$\varphi_k = \frac{2k+1}{2}\langle\varphi_{\lambda_0(g)}, P_k\rangle = \lambda_0^k(g)\,\psi_k(\lambda_0(g)), \quad k = 0, 1, 2, \ldots$$

and Theorem 32 imply that all of the ϕ_k are real and nonzero, and that $\phi_0 = \frac{1}{2}\langle\phi_{\lambda_0(g)}, 1\rangle = 1$,

$$\operatorname{sgn}\varphi_k = \operatorname{sgn}\psi_k(\lambda_0(g)) = (-1)^k, \quad k = 0, 1, 2, \ldots.$$

Moreover,

$$\lim_{k=\infty}\frac{\varphi_{k-1}}{\varphi_k} = \frac{1}{\lambda_0(g)}\lim_{k=\infty}\frac{\psi_{k-1}(\lambda_0(g))}{\psi_k(\lambda_0(g))} = -\alpha,$$

where

$$\alpha = \frac{1+\sqrt{1-\lambda_0^2(g)}}{\lambda_0(g)} > 1.$$

This immediately implies convergence of the series $\sum_{k=0}^{\infty} \phi_k P_k(\mu)$ uniformly with respect to $\mu \in [-1, 1]$.

Solving the latter relation for $\lambda_0(g)$, we obtain

$$\lambda_0(g) = \frac{2\alpha}{\alpha^2+1}.$$

Equations (137) now follow directly from (131) in which we set $\lambda = \lambda_0(g)$.

Thus $\lambda_0(g)$ and g_k, $k = 0, 1, \cdots$, are determined uniquely by specifying $\phi \in \Psi_0$. This means that the class Γ contains exactly one indicatrix g such that $\check{\omega}_0 g = \phi$. This proves Theorem 31. To prove Theorem 33 it remains for us to verify relations (135) and (136).

We set

$$\delta_k = \frac{\varphi_{k-1}}{\varphi_k} + \alpha\left(1 - \frac{1}{2k}\right), \quad k = 1, 2, \ldots.$$

Then $\lim_{k=\infty} \delta_k = 0$, and for all $k \geq 1$ we have

$$-\alpha\left(1 - \frac{1}{2k}\right) + \delta_k \neq 0.$$

By identity transformations we can easily reduce relations (137) to the form

$$h_k = (1 + u_k)\,\delta_k - \frac{1}{\alpha^2}(1 + v_{k+1})\,\delta_{k+1}, \quad k \geqslant 1. \tag{138}$$

Here

$$h_k = \frac{\alpha^2+1}{\alpha} g_k + \frac{1}{2(k+1)(2k+3)} \cdot \frac{1}{\alpha\left(1 - \frac{1}{2k+2}\right) - \delta_{k+1}}; \tag{139}$$

$$u_k = (2k-1)^{-1}, \quad v_k = \left(\frac{\delta_k}{\alpha} + \frac{1}{2k}\right)\left(1 - \frac{1}{2k} - \frac{\delta_k}{\alpha}\right)^{-1}.$$

It is clear that for $k \geq 1$ we have $1 + u_k \neq 0$, $1 + v_k \neq 0$, and that

$$\lim_{k=\infty} u_k = \lim_{k=\infty} v_k = 0.$$

For every $n \geq 1$,

$$\delta_k = \frac{h_k}{1+u_k} + \sum_{\nu=1}^{n} \frac{h_{k+\nu}}{\alpha^{2\nu}} \frac{1}{1+u_k} \prod_{j=1}^{\nu} \frac{1+v_{k+j}}{1+u_{k+j}}$$
$$+ \frac{1}{\alpha^{2n+2}} \frac{1+v_{k+n+1}}{1+u_k} \prod_{j=1}^{n} \frac{1+v_{k+j}}{1+u_{k+j}} \delta_{k+n+1}. \tag{140}$$

This equation can be readily proved by induction over n on the basis of (138). Taking the limit as $n \to \infty$ in (140), we obtain

$$\delta_k = \frac{h_k}{1+u_k} + \sum_{\nu=1}^{\infty} \frac{h_{k+\nu}}{\alpha^{2\nu}} \cdot \frac{1}{1+u_k} \prod_{j=1}^{\nu} \frac{1+v_{k+j}}{1+u_{k+j}}. \tag{141}$$

By (139) we have

$$\sum_{k=1}^{\infty} kh_k^2 < \infty$$

if and only if

$$\sum_{k=0}^{\infty} kg_k^2 < \infty.$$

But the latter condition is fulfilled (compare Theorem 2). This means that

$$\sum_{k=1}^{\infty} kh_k^2 = \mathcal{H} < \infty.$$

Since $\alpha > 1$, there exist a $\theta \in (0, 1)$ and a natural number k_0 such that for $k \geq k_0$ we have

$$|1 + u_k|^{-1} < 2 \text{ and } \left|\alpha^{-2}\frac{1+v_k}{1+u_k}\right| < \theta.$$

Hence for $k \geq k_0$ we find from (141) that

$$|\delta_k| \leqslant 2 \sum_{\nu=0}^{\infty} \theta^{\nu} |h_{k+\nu}|.$$

From this we infer that

$$\sum_{k=k_0}^{\infty} k|\delta_k|^2 \leqslant 4 \sum_{k=1}^{\infty} k\left(\sum_{\nu=0}^{\infty} \theta^{\nu/2}\cdot\theta^{\nu/2}|h_{k+\nu}|\right)^2 \leqslant 4 \sum_{k=1}^{\infty} k\left[\sum_{\nu=0}^{\infty}\theta^{\nu}\right]\left[\sum_{\nu=0}^{\infty}\theta^{\nu}|h_{k+\nu}|^2\right]$$

$$\leqslant \frac{4}{1-\theta}\sum_{\nu=0}^{\infty}\theta^{\nu}\sum_{k=\nu+1}^{\infty} k|h_k|^2 \leqslant \frac{4\mathcal{H}}{(1-\theta)^2} < \infty.$$

This completes the proof of Theorem 33.

Theorem 33 does not describe the class Ψ_0 fully. However, let the function ϕ have the properties listed in Theorem 33: let $\phi(\mu)$ be continuous on $[-1, 1]$ and positive, and let the Fourier coefficients

$$\varphi_k = \frac{2k+1}{2}\langle\varphi, P_k\rangle$$

differ from zero and form an alternating-sign sequence. Let relation (135) be fulfilled for $\alpha > 1$, and let the numbers δ_k have property (136). We set $\lambda_0 = 2\alpha/(\alpha^2 + 1)$ and define the numbers g_k, $k = 0, 1, \cdots$, in accordance with (137). Then

$$\delta_k \to 0, \quad -\alpha\left(1 - \frac{1}{2k}\right) + \delta_k \neq 0,$$

and relations (138), (139) are fulfilled. It is clear that u_k and v_k vanish as $k \to \infty$, so that (138) and condition (136) imply that $\Sigma_{k=1}^{\infty} kh_k^2 < \infty$ and therefore that $\Sigma_{k=0}^{\infty} kg_k^2 < \infty$. This fact enables us to express the function $g \in L_2$ as

$$g = \sum_{k=0}^{\infty} \frac{2k+1}{4\pi} g_k P_k.$$

The above function generally does not belong to the class Γ. However, if we define the kernel $h_0(\mu, \mu')$ by the formula of Lemma 11[iii], the (characteristic-type) equation

$$(1 + \lambda\mu)\varphi(\mu) = \int_{-1}^{1} h_0(\mu, \mu')\varphi(\mu')\,d\mu'$$

is nevertheless fulfilled.

BIBLIOGRAPHY

[1] B. Davison and J. B. Sykes, *Neutron transport theory*, Clarendon Press, Oxford, 1957; Russian transl., Atomizdat, Moscow, 1960. MR 20 #2217.

[2] E. Hopf, *Mathematical problems of radiative equilibrium*, Cambridge Tracts in Math. and Math. Phys., no. 31, Cambridge Univ. Press, New York, 1934.

[3] G. Placzek and W. Seidel, *Milne's problem in transport theory*, Phys. Rev. (2) 72 (1947), 550–555. MR 9, 147.

[4] V. Kourganoff, *Basic methods in transfer problems. Radiative equilibrium and neutron diffusion*, Clarendon Press, Oxford, 1952. MR 14, 879.

[5] I. W. Busbridge, *The mathematics of radiative transfer*, Cambridge Univ. Press, New York, 1960.

[6] S. Chandrasekhar, *The transfer of radiant energy*, Clarendon Press, Oxford, 1950; Russian transl., IL, Moscow, 1953. MR 13, 136.

[7] J. C. Mark, *Milne's problem for anisotropic scattering*, National Research Council of Canada, Division of Atomic Energy, Document no. MT–26, 1943. MR 9, 444.

[8] V. A. Fok, *Sur certaines equations intégrales de physique mathématique*, Mat. Sb. 14 (56) (1944), 3–50. (Russian) MR 6, 272.

[9] I. C. Gohberg and M. G. Kreĭn, *Systems of integral equations on a half line with kernels depending on the difference of arguments*, Uspehi Mat. Nauk 13 (1958), no. 2 (80), 3–72; English transl., Amer. Math. Soc. Transl. (2) 14 (1960), 217–287. MR 21 #1506; MR 22 #3954.

[10] T. A. Germogenova, *Bounded solutions of an inhomogeneous integral equation given on a semi-infinite interval and having its kernel dependent on the difference of the arguments*, Dokl. Akad. Nauk SSSR 115 (1957), 23–26. (Russian) MR 21 #2884.

[11] I. Kuščer, *Milne's problem for anisotropic scattering*, J. Mathematical Phys. 34 (1956), 256–266. MR 18, 705.

[12] L. M. Romanova, "The Milne problem for a half-space with anisotropic scattering and capture of neutrons," *Collection: Some mathematical problems of neutron physics*, Moscow Univ. Press, Moscow, 1960.

[13] P. Bouvier, *Une extension particulière de la méthode de Wiener-Hopf*, Arch. Sci. Soc. Phys. Hist. Nat. Genève 8 (1955), 87–93. MR 16, 1162.

[14] G. A. Bat' and D. F. Zareckiĭ, "Solving the generalized Milne problem," *Collection: Reactor design and reactor theory*, Papers of the Soviet Delegation to the Geneva Conference, 1955, Izdat. Akad. Nauk SSSR, Moscow, 1955, p. 294

[15] J. R. Mika, *Neutron transport with anisotropic scattering*, Nuclear Sci. Engineering 11 (1961), 415. Physics Abstracts 65 #5663.

[16] A. Leonard and T. W. Mullikin, *Spectral analysis of the anisotropic neutron transport kernel in slab geometry with applications*, J. Mathematical Phys. 5 (1964), 399–411. MR 30 #926.

[17] I. B. Russman, *Reducing the problem of radiative transfer in a half-space to a singular equation*, Ž. Vyčisl. Mat. i Mat. Fiz. 4 (1964), no. 6, 1122–1126 = U.S.S.R Comput. Math. and Math. Phys. 4 (1964), no. 6, 206–212.

[18] V. A. Ambarcumjan, *A new method for the calculus of the light dispersion in a turbid medium*, Izv. Akad. Nauk SSSR Ser. Geogr. Geofiz. 1942, 97–103. (Russian) MR 4, 92.

[19] I. Kuščer, *Diffusion of radiation in an infinite scattering medium*, Bull. Sci. Conseil Acad. RSF Yougoslav. 2 (1954), 15.

[20] V. V. Sobolev, *Radiative energy transfer in stellar and planetary atmospheres*, Fizmatgiz, Moscow, 1956. (Russian)

[21] M. V. Maslennikov, *Mathematical problems of the theory of neutron retardation and diffusion*, Thesis, Moscow, 1958.

[22] ———, *Milne's problem with an arbitrary indicatrix*, Dokl. Akad. Nauk SSSR 118 (1958), 895–898. (Russian) MR 20 #3713.

[23] ———, *A method of solving the characteristic equation of radiation transfer theory*, Ž. Vyčisl. Mat. i Mat. Fiz. 1 (1961), 255–266 = U.S.S.R Comput. Math. and Math. Phys. 1 (1961), 289–303.

[24] ———, *On a converse problem in the theory of radiation passage through matter*, Dokl. Akad. Nauk SSSR 145 (1962), 1019–1021 = Soviet Phys. Dokl. 7 (1963), 688–689. MR 26 #7309.

[25] ———, *Uniqueness of the solution of the inverse problem of the asymptotic theory of radiative transfer*, Ž. Vyčisl. Mat. i Mat. Fiz. 2 (1962), 1044–1053 = U.S.S.R Comput. Math. and Math. Phys. 2 (1962), 1247–1260. MR 26 #5968.

[26] M. V. Maslennikov and T. A. Suškevič, *Asymptotic properties of the solution of the characteristic equation of the transport theory of radiation in strongly absorbent media*, Ž. Vyčisl. Mat. i Mat. Fiz. 4 (1964), 23–34 = U.S.S.R. Comput. Math. and Math. Phys. 4 (1964), no. 1, 29–46. MR 29 #1876.

[27] E. W. Hobson, *The theory of spherical and ellipsoidal harmonics*, Chelsea, New York, 1955. MR 16, 356.

[28] A. Sommerfeld, *Partielle Differentialgleichungen der Physik*, Vorlesungen über theoretische Physik, Band VI, Akademische Verlagsgesellschaft Geest & Portig, Leipzig, 1947. MR 10, 195.

[29] V. S. Vladimirov, *Mathematical problems of the uniform-speed of transport*, Trudy Mat. Inst. Steklov. 61 (1961). (Russian) MR 27 #6579.

[30] P. Funk, *Beiträge zur Theorie der Kugelfunktionen*, Math. Ann. 77 (1915), 136–152.

[31] E. Hecke, *Über orthogonal-invariante Integralgleichungen*, Math. Ann. 78 (1918), 398–404.

[32] V. S. Vladimirov, *An integral equation related to spherical functions*, Naučn. Dokl. Vysš. Školy. Fiz.-Mat. Nauki 1958, no. 6, 142.

[33] L. V. Kantorovič and G. P. Akilov, *Functional analysis in normed spaces*, Fizmatgiz, Moscow, 1959; English transl., Internat. Series of Monographs in Pure and Appl. Math., vol. 46, Macmillan, New York, 1964. MR 22 #9837; MR 35 #4699.

[34] M. G. Kreĭn and M. A. Rutman, *Linear operators leaving invariant a cone in a Banach space*, Uspehi Mat. Nauk 3 (1948), no. 1 (23), 3–95; English transl., Amer. Math. Soc. Transl. (1) 10 (1962), 199–325. MR 10, 256; MR 12, 341.

[35] E. Hille and R. S. Phillips, *Functional analysis and semi-groups*, Amer. Math. Soc. Colloq. Publ., vol. 31, Amer. Math. Soc., Providence, R. I., 1957; Russian transl., IL, Moscow, 1962. MR 19, 664.

[36] B. Sz.-Nagy, *Perturbations des transformations linéaires fermées*, Acta Sci. Math. Szeged 14 (1951), 125–137. MR 13, 849.

[37] I. M. Gradšteĭn and I. M. Ryžik, *Tables of integrals, sums, series, and products*, 4th ed., Moscow, 1963; English transl., Academic Press, New York, 1965. MR 28 #5198; MR 33 #5952.

[38] I. G. Petrovskiĭ, *Lectures on the theory of integral equations*, "Nauka", Moscow, 1966; English transl., Graylock Press, Rochester, New York, 1957. MR 19, 42.

[39] E. C. Titchmarsh, *Introduction to the theory of Fourier integrals*, Clarendon Press, Oxford, 1937; Russian transl., GITTL, Moscow, 1948.

[40] O. Perron, *Die Lehre von Kettenbrüchen*, Teubner, Berlin, 1929.

[41] E. H. Bareiss, *A spectral theory for the stationary transport operator in slab geometry*, J. Math. Anal. Appl. 13 (1966), 53–92. MR 32 #7164.

[42] M. G. Kuz'mina and M. V. Maslennikov, *Dependence of the asymptotic characteristics of the solution of the Milne problem on the scattering indicatrix*, Ž. Vyčisl. Mat. i Mat. Fiz. 8 (1968), 573–589 = U.S.S.R Comput. Math. and Math. Phys. 8 (1968), no. 3 (to appear).

[43] M. V. Maslennikov, *The characteristic equation of the theory of radiation transfer*, Dokl. Akad. Nauk SSSR 168 (1966), 747–750 = Soviet Math. Dokl. 7 (1966), 723–726. MR 33 #4638.

[44] ———, *The Milne problem with an arbitrary indicatrix and in the case of azimuthal nonhomogeneity of solution*, Dokl. Akad. Nauk SSSR 168 (1966), 1001–1004 = Soviet Math. Dokl. 7 (1966), 768–771. MR 33 #4639.

[45] G. I. Marčuk, *Numerical methods for nuclear reactor calculations*, Atomizdat, Moscow, 1958; English transl., Consultants Bureau, New York and Chapman & Hall, London, 1959. MR 21 #3109.

Translated by:
Andrew Yablonsky